Shapes and Spins of Near-Earth Asteroids

Michael W. Busch

DISSERTATION.COM

Boca Raton

Shapes and Spins of Near-Earth Asteroids

Dissertation.com
Boca Raton, Florida
USA • 2010

ISBN-10: 1-59942-322-7
ISBN-13: 978-1-59942-322-7

I dedicate this thesis to Steven J. Ostro. Steve inspired me to come to Caltech, and was an outstanding mentor and advisor. We all miss him greatly.

Acknowledgments

I must thank Shri Kulkarni for equally valuable guidance, and for accepting a thesis project so divergent from the rest of his work. The other members of my thesis committee, Oded Aharonson, Geoff Blake, and Dave Stevenson, also provided very valuable feedback.

Lance Benner and Mike Nolan have been great teachers of asteroid radar astronomy. Jon Giorgini deserves special thanks for producing the ephemeris programs `horizons` and `OSOD`, essential for radar observations. Christopher Magri has developed the code used for asteroid shape modeling from radar data. Daniel Scheeres' dynamical work has been invaluable in understanding the implications of my asteroid shape models. Walter Brisken and Adam Deller developed the `NRAO-DiFX` software correlator used during the VLBA observations and Craig Walker modified the VLBA tracking routines to accommodate targets inside the solar system. The staffs of the NRAO at Socorro, the Goldstone Solar System Radar, and the Arecibo Observatory made all of this work possible.

My fellow students and my parents deserve great gratitude for their moral support and for agreeing to read through this entire manuscript.

Finally, I must thank Paul Young, Lowell Wood, and the directors of the Fannie and John Hertz Foundation, for their generosity in funding my work for the past three years.

Abstract

Asteroids are diverse and numerous solar system objects, from the large number of objects in the main asteroid belt to the relatively small near-Earth population. Understanding their physical properties is essential to understanding the evolution of the solar system, and asteroid morphology is a complex field in its own right. The histories of individual asteroids, and particularly near-Earth objects, reflect continuous interaction among their shapes, rotation states, and orbits due to the effects of radiation pressure.

Radar astronomy has provided detailed information on the orbits, sizes, shapes, rotation states, and composition of many asteroids. To improve the capabilities of asteroid radar observations, I have developed the technique of radar speckle tracking. The echoes from different points on the surface of a radar target interfere with each other, producing a pattern of bright and dark speckles across the surface of the Earth. Using radio astronomy techniques, I track the motion of speckles between several ground stations during a radar experiment to accurately determine the rotation state of the target. Speckle tracking is a powerful tool both to determine the orbital evolution of near-Earth asteroids, particularly potential Earth impactors, and to survey the overall physical properties of the asteroid population.

In addition, I have studied applying the techniques of adaptive optics and radio interferometry to asteroid science. These will become more useful with the next generation of asteroid-detecting surveys and the construction of large sub-millimeter interferometers. Interferometry in particular will soon be able to survey the entire asteroid belt.

Contents

List of Figures viii

Nomenclature ix

1. Introduction 1

2. Overview of Asteroids in the Solar System 2
   a. Asteroid populations 2
      i. The main belt 2
      ii. The near-Earth asteroids 2
   b. Asteroid morphology 4
      i. Shapes and composition 4
      ii. The Yarkovksy effect 6
      iii. YORP and reconfiguration 8

3. Asteroid Radar Astronomy 10
   a. Transmitters and Receivers 10
   b. Radar echoes 11
      i. Polarization ratio and surface roughness 12
      ii. Radar albedo and near-surface bulk density 13
      iii. Phobos and Deimos 15
   c. Doppler resolution and range coding 18
   d. Shape modeling 19
      i. SHAPE software 20
      ii. 1992 SK 23
      iii. 1998 WT24 26
      iv. 1950 DA 30
   e. Limitations of shape modeling: pole direction ambiguities ... 33
      i. 1950 DA in 2880 33
      ii. 2008 EV5 35

4. Ways to Determine Asteroid Pole Directions 40
   a. Adaptive optics 40
   b. Radar interferometric imaging 42
      i. VLBI arrays 42
      ii. Definition of visibility 43
      iii. Image reconstruction from visibility data 46
      iv. Limitations of synthesis imaging 46
      v. Conditions for radar-interferometric imaging 47
      vi. Radar speckles and interferometry 51

    c. Radar speckle patterns 52
        i. Speckle patterns and pole directions 52
        ii. Structure of a speckle pattern 55
        iii. Capabilities of speckle tracking 57

5. **Implementing Radar Speckle Tracking** **58**
    a. Radar speckle processing for the Very Long Baseline Array 58
        i. Computation of cross-correlation 58
        ii. The software correlator 59
        iii. Changing t_{lag} 60
        iv. Station selection for speckle observations 61
    b. Application to a near-Earth asteroid: 2008 EV5 63
    c. Other applications of speckle tracking 64

6. **Future Possibilities in Asteroid Science** **66**
    a. Effect of PanSTARRS, LSST, and other surveys.................... 66
    b. Future radar facilities 66
    c. ALMA, CARMA, and thermal radiation 67
    d. A final thought 71

References **72**

Appendix 1: Delay-Doppler radar images used in shape modeling **80**
    a. 1998 WT24 80
    b. 1950 DA 87
    c. 2008 EV5 92

Appendix 2: Narrow-band software correlator **96**
    a. Input files 96
    b. Output 98
    c. Source code 99

List of Figures

2.1. View of the inner solar system. Courtesy Minor Planet Center. 3

2.2. Shapes of various asteroids. Figure from Ostro et al. 2010. 5

2.3. Near-Earth asteroid Itokawa. Figure from Fujiwara et al. 2006. 6

2.4. Yarkovsky and YORP forces acting on an object. 7

3.1.   a. Arecibo Observatory. 11

        b. Goldstone 70-m antenna. 12

3.2. Relationships between bulk density and Fresnel reflection coefficient. 14

3.3. Echo power spectra of Phobos and Deimos. 15

3.4. Schematic of Doppler-only radar resolution. 17

3.5. Delay-Doppler images of the asteroid 1992 SK. 18

3.6. Nominal retrograde model of 2008 EV5 and two alternate models. 22

3.7. Selected delay-Doppler images of 2008 EV5 and corresponding fits. 23

3.8. Principal axis projections of the 1992 SK model. 24

3.9. 1992 SK model shaded for gravitational slope. 25

3.10. Selected delay-Doppler images of 1998 WT24. 27

3.11. 1998 WT24 model in principle axis projection. 29

3.12. Selected delay-Doppler images of 1950 DA. 31

3.13. Principal-axis projections of the models of 1950 DA. 32

3.14. Predicted position of 1950 DA. Plot courtesy Jon D. Giorgini. 34

3.15. Selected delay-Doppler images of 2008 EV5. 36

3.16. Nominal prograde and retrograde models of 2008 EV5. 38

4.1. Keck adaptive optics image of 2004 XP14. 41

4.2. A representative interferometric array: the VLBA. 42

4.3. Interferometer response as a function of source and baseline. 45

4.4. Simulated VLBA observations at 2380 and 8560 MHz. 48

4.5. Potential radar interferometric targets at Arecibo and Goldstone. 50

4.6. Radar speckle pattern of 2008 EV5. 51

4.7. Schematic of a radar speckle pattern. 53

4.8. Power spectrum of 2008 EV5 radar speckles and a simulated sphere. 56

5.1. Radar echo of EV5 processed with my software correlator. 60

5.2. Comparison between my narrow-band correlator and DiFX. 61

5.3. Asteroid radar targets suitable for speckle tracking. 62

5.4. Speckle tracking of 2008 EV5, showing retrograde rotation. 64

5.5. Principal axis views of the best current EV5 shape model. 65

6.1. Simulated ALMA image of the asteroid 216 Kleopatra. 69

Nomenclature

a	Radar albedo.
$A_{ij}(\sigma)$	Product of the antenna beam pattern (gain as a function of σ) for two antennas i and j.
A_r	Effective antenna collecting area.
ALMA	Atacama Large Millimeter Array, currently under construction.
$\vec{b}_{ij}$	Antenna baseline: vector separation of antennas i and j perpendicular to the line-of-sight.
b	Backscatter gain of a surface for radar reflections.
CARMA	Combined Array for Research in Millimeter-wave Astronomy
D	Distance between two antennas in an array (baseline length).
D_{max}	Maximum distance between stations in an array.
$D_{nearfield}$	Near-field distance of an array.
D_{target}	Distance to target.
d	Diameter of target.
$E(\sigma)$	Electric field radiated from a point on the sky.
$F()$	Denotes Fourier transform.
GBT	Green Bank Telescope
G_i	Antenna gain amplitude.
$G_i(\sigma)$	Antenna gain pattern: gain as a function of sky position.
$G_{ij}(\sigma)$	Product of antenna gain patterns of two stations.
g_i	Antenna gain, including both amplitude and phase. Determined by antenna shape and size, observation frequency, antenna pointing, and position on the sky.
$I(\sigma)$	Brightness as a function of plane-of-sky position.
LSST	Large Synoptic Survey Telescope, currently under construction.
N	Number of antennas/stations in an array.
OC	Opposite-sense circular polarization as transmitted by a radar.
P	Rotation period of an object.
P_t	Radar transmit power.
P_r	Radar receive power.
Pan-STARRS	Panoramic Survey Telescope And Rapid Response System
R	Fresnel reflection coefficient.
r	Object radius.
$\vec{R}(\sigma)$	Radius vector to one location on the sky.
$\vec{r}_i$	Vector position of one station in an array, usually measured relative to the Earth's center of mass.
SC	The same-sense circular polarization as transmitted by a radar.
SC/OC	Ratio of echo power in same-sense and opposite-sense circular polarizations.
T_{noise}	Noise temperature of receiver.
t_{int}	Integration time of a measurement
t_{lag}	Time lag in a signal (particularly radar speckles) between two stations.
Δt_{lag}	Uncertainty in t_{lag}.

V_i	Receiver voltage measured at a station.
V_{ij}	Visibility = complex cross-correlation between two stations.
VLA	Very Large Array
VLBA	Very Long Baseline Array
α	Angle between a baseline and a radar target's spin vector projected to the plane of the sky.
λ	Wavelength.
σ	Position on the sky, typically measured in RA-Dec.
ρ	Bulk density of material.
θ	Sub-Earth latitude on a radar target.
ν	Frequency.
$\Delta\nu$	Denotes bandwidths and frequency resolution.

Note on asteroid names:

Asteroids and other minor planets are assigned an alphanumeric code in order of their discovery. This name takes the form <year of discovery> <letter denoting 2-week period of that year><letter denoting the object's order of discovery in that two-week block><optional number>. Thus, '1950 DA' was discovered in 1950, in the second half of February, and was the first object discovered in that interval, and '1992 SK' was the 11th object discovered in the 'S'-period of 1992. When more than 26 objects are discovered in a two week period, the second letter is reused, followed by a number giving the number of times through the alphabet the naming has gone. For instance, the discovery of the object '2008 EZ' in the first half of March 2008 was followed by the discovery of the object '2008 EA2', which was followed by '2008 EB2'. '2008 EV5' was the 126th object discovered in that period. Additionally, if what initially appears to be a new object is determined to be a recovery of an object that had been lost, the older designation is used – thus there are gaps in the list.

Once the orbit solution for an object is refined to a specified precision, it is assigned a catalog number. Catalog numbers are given in order of orbit refinement, rather than of discovery: e.g., 29075 (1950 DA) is followed by 29076 (1972 TR8). After having a catalog number, the object can officially named: 99942 (2004 MN4) was named Apophis. Objects assigned official names are no longer referred to by alphanumerics, and may be mentioned with or without catalog numbers. However, since the majority of asteroids are not named, alphanumerics and truncations of them are used extensively in the asteroid literature and in this thesis.

1. Introduction

In this thesis, I present my work on the general problem of understanding the physical properties of asteroids. Asteroids are diverse and numerous solar system objects, ranging from the large number of objects in the main belt to the relatively small near-Earth population. Understanding their physical properties is essential to understanding the evolution of the solar system and asteroid morphology is a complex field in its own right. The histories of individual asteroids, particularly near-Earth objects, reflect continuous interaction between their shapes, rotation states, and orbits due to the effects of radiation pressure. I discuss these interactions and the asteroid populations in detail in Chap. 2.

The bulk of my work has been radar shape modeling and speckle tracking as applied to near-Earth asteroids and these topics occupy the appropriate fraction of the material. A part of Chap. 4 describes the results of adaptive optics observations of near-Earth and main-belt asteroids. Please consider this section an interesting diversion from the main science results.

Prior to developing radar speckle tracking, I studied the moons of Mars and several near-Earth asteroids (1950 DA, 1992 SK, 1998 WT24, and 2008 EV5) with conventional radar techniques, using archival and new data from both the Arecibo and Goldstone radar observatories. The results of this work are presented in Chap. 3, and serve as a description of the capabilities of radar astronomy and a demonstration of the need to understand asteroid pole directions. Chapter 4 describes different methods of measuring asteroid rotation states: adaptive optics, radar interferometric imaging, radar speckle tracking, and the capabilities and limitations of each. In Chap. 5, I describe my implementation of radar speckle tracking (source code and documentation is included as Appendix 2), its application to observations of the near-Earth asteroid 2008 EV5 with Arecibo and the Very Long Baseline Array, and some future applications.

I conclude with some longer-term research possibilities: the pending explosion in the number of known asteroids with the next generation of optical surveys, and the potential of submillimeter interferometry to image asteroids by detecting their thermal emission rather than a radar echo (Chap. 6).

2. Overview of Asteroids in the Solar System

Asteroids are small solar system objects, generally defined as being too small to gravitationally dominate a significant region of space and composed primarily of non-icy materials. They range from sub-meter boulders to the 800-km Ceres and from objects that cross the orbit of Mercury to the Trojan asteroids co-orbiting with Jupiter. My work has been driven by a desire to understand their physical properties, particularly shapes and spin states. To justify this as a worthy field of study, I will describe the overall distribution and properties of asteroids in the solar system.

2.a. Asteroid Populations

2.a.i. The main belt

The vast majority of asteroids, by both number and mass, reside in the main asteroid belt between the orbits of Jupiter and Mars. The belt originated dynamically, as the region in the proto-planetary disc where the planetesimals were perturbed by Jupiter to have random velocities too high to accrete into a small number of objects (Petit et al. 2001). During planetary migration, a significant number of ice-rich objects from the outer solar system may have been implanted into the main belt as a result of dynamical interactions with Jupiter (Levison et al. 2009). Since that time, the belt has developed an intricate structure (Fig. 2.1). There has been a gradual decrease in the mass of the belt as objects migrate onto orbits that are perturbed by the gas giants or are collisionally ground into dust. Some of the migrating objects enter the inner solar system, forming the near-Earth population (red circles in Fig. 2.1), which I discuss below.

In the asteroid belt, three subpopulations are particularly significant. The Jupiter Trojans (dark blue points in Fig. 2.1) reside close to the Sun-Jupiter Lagrange points. The Trojans are dynamically stable and the physically coldest asteroids. They are relatively ice-rich and chemically similar to the Centaur objects at larger semimajor axis (Barucci et al. 2002). The Hilda asteroids are in a 3:2 mean-motion resonance with Jupiter, and seen as the threefold pattern of the outer main belt in Fig. 2.1. This resonance is stable, in contrast to the mean-motion and secular resonances at lower semi-major axis, where the asteroids are expelled from the main belt, producing the drops in asteroid number as a function of semimajor axis known as the Kirkwood Gaps (Kirkwood 1866). Finally, one-third of the asteroids in the main belt are members of collisional families, the remnants of larger objects disrupted relatively recently in the history of the solar system. All but the largest asteroids have experienced disruptive collisions; the size distribution of asteroids smaller than 100 km is well approximated by a collisional cascade power law (Bottke et al. 2005). Despite this violent history, there is an overall gradient in composition across the main belt, particularly in the outer belt where subsurface ice becomes stable.

2.a.ii. The near-Earth asteroids

Near-Earth asteroids are more chaotic than their progenitors in the main belt, in terms of both their orbits and their composition. As seen in Fig. 2.1, most asteroids with perihelion <1.3 AU (the conventional boundary of the near-Earth population) are on highly eccentric orbits. Objects at low inclination have frequent close flybys of the inner planets, particularly Earth, Venus, and Mars, and are strongly perturbed by Jupiter and main belt asteroids when they are close to aphelion. Due to such frequent perturbations, a typical near-Earth asteroid has a dynamical lifetime of <100 Myr. None of the

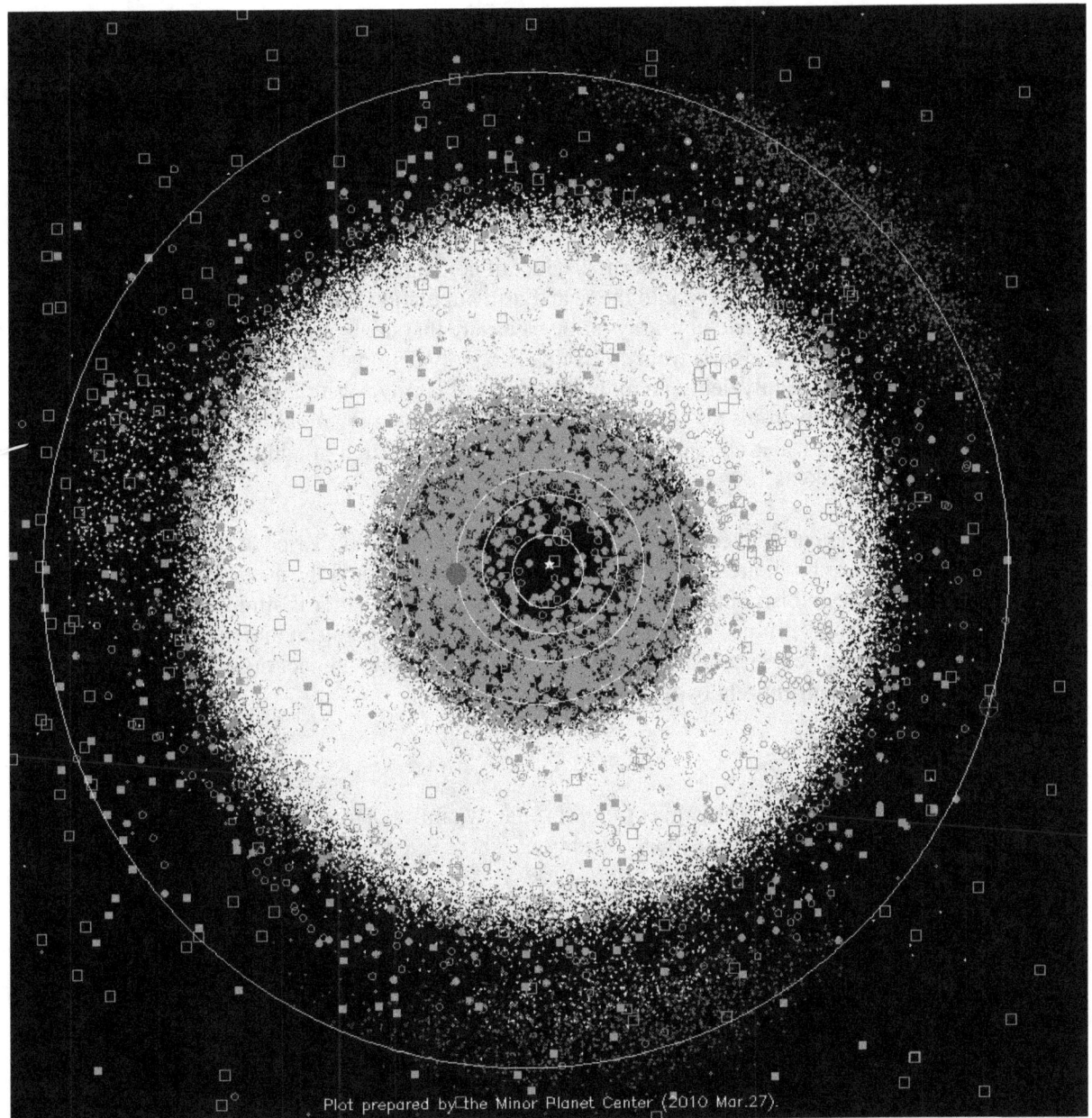

Fig. 2.1. *Projected view of the inner solar system from ecliptic north at 2010 March 27, 00:00 UTC, illustrating the different asteroid populations. This plot contains a total of roughly 250000 objects. Light blue squares are comets; dark blue points are Jupiter Trojans (Jupiter's orbit is the outermost blue curve, the planet itself is at lower right). Green points indicate the main belt asteroids, and red circles the near-Earth population. Earth is to the left of the Sun. The majority of comets that appear to be well within the orbit of Jupiter are high-inclination objects at larger distances from the Sun. A note on scale: the orbit of Jupiter has a radius of 5.2 astronomical units (780 million kilometers); the largest asteroids are approximately 1/2000 the width of a pixel. Plot courtesy Gareth Williams, Minor Planet Center (http://www.cfa.harvard.edu/iau/lists/InnerPlot.html).*

current near-Earth objects is primordial, and the overall number of objects fluctuates on tens-of-Myr timescales, with increases following large family-forming collisions in the main belt. The family-forming collisions produce a population of objects that migrate due to perturbations and non-gravitational effects – particularly net radiation pressure from non-isotropic thermal emission (the Yarkovsky effect, see Sec. 2.b). When migration puts the objects into one of the unstable resonances with Jupiter and Saturn, their eccentricity can increase until they become near-Earth objects (Bottke et al. 2007).

In addition to objects that are the recent product of collisions, some primordial objects migrate into Jupiter/Saturn resonance and move into the near-Earth population. Regardless of their origin, perturbations (both gravitational and non-gravitational) ensure that none of the near-Earth objects forms a long-lived collisional family. Debris from collisions is rapidly disrupted and spread out. Meteor streams are the closest to an equivalent of a main-belt asteroid family. With low ejection velocity from their parent comets, the meteoroids do not spread in orbital elements very quickly. Even so, they disperse on 1000-year timescales and only are maintained by the ejection of additional material from their parent bodies (Jenniskens 1994).

There are four potential fates for any given near-Earth asteroid (O'Brien & Greenberg 2005). Most either hit a planet or hit or are hit by another asteroid. The former is the asteroid impact hazard, when the planet is Earth, and is of the most interest to the public. The latter is simply the collisional cascade, and produces smaller objects that are even more affected by non-gravitational forces. Asteroids that do not suffer collisions either reach such high eccentricity that they impact the Sun or are scattered out of the inner solar system entirely by interactions with Jupiter.

2.b. Asteroid Morphology

I have described asteroids in terms of their orbital dynamics. However, asteroids are not point masses or simply the remnants of collisional disruption. There is a huge diversity of asteroid shapes, sizes, compositions, and rotation states.

2.b.i. Shapes and composition

For the largest asteroids – Ceres, Pallas, Vesta, and Hygiea – gravity dominates over material strength and their shapes are relatively close to hydrostatic equilibrium. At smaller sizes (200-300 km), rock's compressive strength becomes comparable to gravitational pressure and many asteroids this size are notably non-spherical (such as Iris and Kleopatra, Fig. 2.2). For still smaller objects – including Ganymed and Eros, the largest two near-Earth asteroids – the fracture and granular properties of the asteroids' interiors determine their overall shapes and response to impacts. These objects are generally referred to as rubble piles (Fig. 2.3, Harris 1996). Still smaller asteroids, <200 m, can consist of single unfractured blocks and have significant tensile strength.

Large asteroids differentiated into metallic cores and silicate mantles early in the history of the solar system, due to heat produced from the decay of short-lived radioisotopes – particularly ^{26}Al and ^{60}Fe – and developed complex mineralogies (e.g. Day et al. 2009). Many of these objects have since been disrupted by collisions, leaving a few large nickel-iron bodies such as the bilobate asteroid Kleopatra (Ostro et al. 2000) and many small objects made of nickel-iron, silicate phases produced at high temperature, or a mixture of the two. Asteroids that never differentiated contain carbonaceous

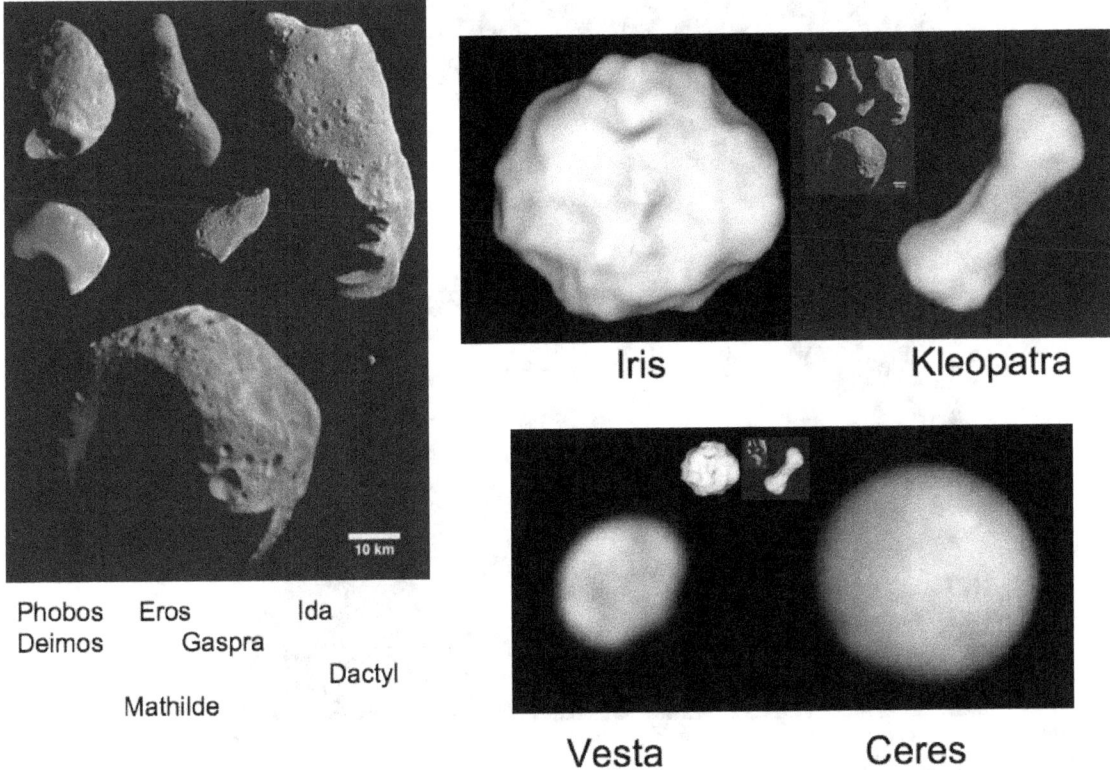

Iris   Kleopatra

Phobos   Eros   Ida
Deimos   Gaspra
    Dactyl
Mathilde

Vesta   Ceres

Fig. 2.2. Shapes of various asteroids, illustrating the transition from gravity-dominated (Ceres, Vesta, Iris) to strength-dominated shapes (Mathilde and smaller). The left panel is a zoomed-in version of the inset in the upper right panel, which in turn is a zoom-in of the inset in the lower right. The bilobate asteroid Kleopatra may be an anomaly (see text). Phobos and Deimos are included here as well. Eros is the second-largest near-Earth asteroid. Itokawa (Fig. 2.3) and the objects I discuss in Chapters 3, 4, and 5 are all comparable in size to or smaller than Dactyl. Figure from Ostro et al. 2010.

material mixed with silicate and metallic grains (e.g. the carbonaceous chondrite meteorites, some of which have been chemically unaltered for 4.567 Gyr, Baker et al. 2005). In the outer asteroid belt and Trojan populations, low bulk densities (Marchis et al. 2006a) and occasional outgassing imply that a significant fraction of the mass is water ice rather than rock or metal ("main-belt comets", Hsieh & Jewitt 2006).

To date, most of our knowledge of asteroid composition comes from meteorites analyzed in the laboratory (e.g. Burbine et al. 2002). Optical and near-infrared spectroscopy provide information on the composition of asteroids currently in space, by comparing the spectra of different asteroids to each other to identify common spectral features – spectral types – and to laboratory spectra of meteorites to relate spectral types to the presence of particular minerals. In situ observations by spacecraft have been helpful (Veverka et al. 2000), as will sample return such as that attempted by the Hayabusa spacecraft from the asteroid Itokawa (Fujiwara et al. 2006).

There are several different systems of spectral typing in use, and they are constantly being adjusted (e.g. Bus et al. 2002). Three major spectral types are of particular interest, and are mentioned throughout this thesis: the S-type objects, which have spectra dominated by silicate absorption features;

Fig. 2.3. The near-Earth asteroid Itokawa, as imaged by the Hayabusa spacecraft. Itokawa is 535 m long, and shows a rubble-pile structure. Figure from Fujiwara et al. 2006.

the C-type objects, inferred to be dominated by carbonaceous material and hydrated silicates; and the X-type objects, which have spectra that are nearly featureless in the optical and near-infrared other than a slight slope to the red. The X-types are sub-divided into E-type objects, which have high optical albedo and are dominated by the high-formation-temperature silicate mineral enstatite; M-types, which have intermediate optical albedo and can be either high-density nickel-iron or low-density hydrated silicates; and P-types, which have low optical albedo and are mixtures of hydrated silicates and carbonaceous materials.

Spectroscopy and photometry alone are unable to distinguish between the different X-type objects; this requires estimates of the object's albedo and bulk density. Even an albedo estimate by itself is sometimes insufficient. Prior to the discovery of the satellite of the non-metallic M-type asteroid Kalliope, all M-types were believed to be metallic, but Kalliope's low mass requires that it be dominated by silicates (Margot & Brown 2003). As I describe in Sec. 3.d, radar astronomy provides information on asteroid composition, by allowing albedo estimates and approximate measurements of the near-surface bulk densities of target objects, even those without satellites. Radar observations have been essential in determining the composition of many asteroids (Benner et al. 2008).

2.b.ii. The Yarkovsky effect

I have mentioned the importance of non-gravitational forces in driving the migration of small asteroids within the main belt and from the main belt to the near-Earth population. The general process at work here is called the Yarkovsky effect, and was first described circa 1900 by Ivan Osipovich Yarkovsky (Öpik 1951).

Consider an object large enough that thermal conduction does not equilibrate the temperature on the timescale of a rotation (in practice, any asteroid larger than a few meters). Then the temperature will vary across the surface, due to the balance between the absorbed solar flux and thermal re-radiation. Since there is some thermal inertia, the hottest point on the surface will be in the afternoon region of the object and there is a net imbalance in the radiation pressure. More photons are radiated from the hotter region and the object experiences a net force in the opposite direction (Fig. 2.4). To determine the true Yarkovsky force on an object requires a model of its shape, pole direction, rotation rate, surface thermal properties, and orbit to determine the solar insolation and the resulting pattern of thermal re-emission. To convert from the Yarkovsky force to the object's acceleration requires an estimate of the object's mass (Giorgini et al. 2002, Chesley et al. 2003, Bottke et al. 2005).

The absorbed and reflected portions of the incident sunlight also carry a certain amount of momentum. This radiation pressure is directed almost radially outward from the Sun and therefore acts as a decrease in the effective solar gravity. It is also determined by the asteroid's shape, pole direction, and mass, but is generally considered separately from the Yarkovsky effect (Bottke et al. 2005).

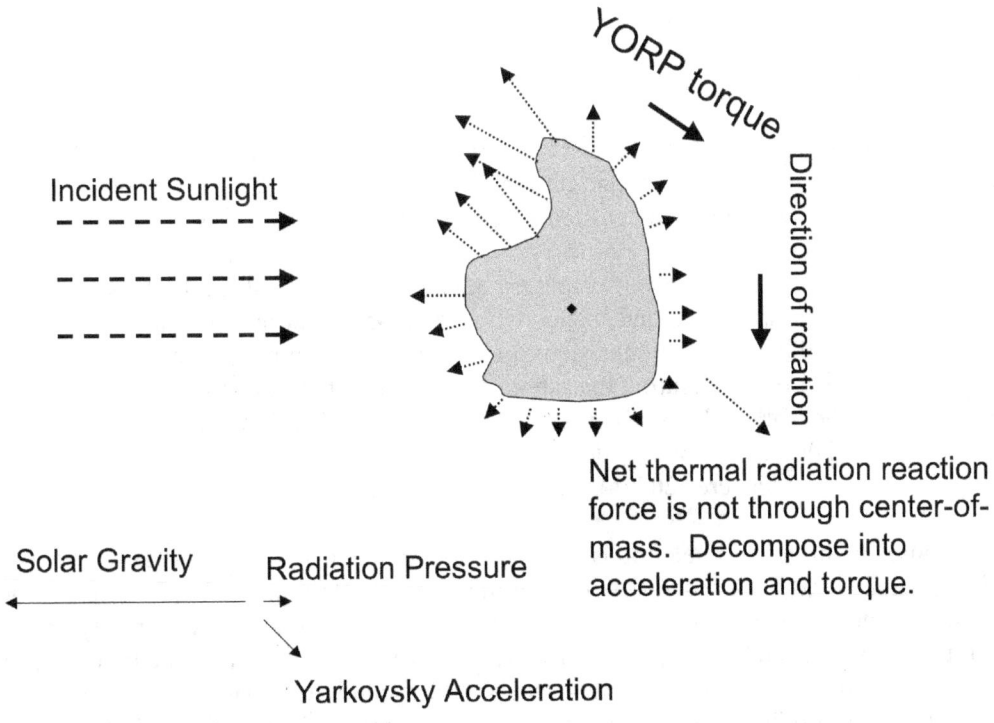

Figure 2.4. Yarkovsky and YORP forces acting on an object, showing the importance of the rotation state, shape, and thermal inertia. The magnitudes of the Yarkovsky and radiation pressure accelerations relative to the Sun's gravity are greatly exaggerated, as is the magnitude of the YORP torque compared to the current rotation.

Yarkovsky migration causes dramatic effects on the trajectories of individual objects. For a typical 1-km near-Earth asteroid, the Yarkovsky force is of order 1 N, producing an acceleration of order 10^{-12} m/s^2. Over 100 Myr, this can accumulate to a few km/s – a significant fraction of the asteroid's total orbital velocity and sufficient to produce close planetary encounters when none was possible before. In the main asteroid belt, sunlight is weaker, and a similar change in velocity would take ~1 Gyr. Over the age of the solar system, Yarkovsky has moved 10-km objects from stable main belt orbits into unstable resonance and in turn into the near-Earth population.

To predict Earth impacts with 'certainty', the uncertainty in a potential impactor's position must be smaller than an Earth diameter. The Yarkovsky acceleration on objects in the 100 m -1 km size range (those with the potential for regional to global destruction on millennium timescales) accumulates to a change in position of more than an Earth radius within decades. Understanding asteroids' physical properties – shape, spin state, and mass – is essential to predicting Earth impacts centuries in the future (Giorgini et al. 2002, 2008).

Working backward in time rather than forward, models of Yarkovksy are essential for accurately reconstructing the progenitors of asteroid families: smaller family members migrate away from their initial orbits. With proper modeling, the asteroid belt can be studied forensically, and collisions between even relatively small asteroids dated to within a few hundred thousand years (Nesvorny et al. 2006). Yarkovsky may not have been important during planetary formation, because the optically thick protoplanetary disc blocked and redistributed sunlight. However, to understand what structures in the asteroid belt result from processes in the early solar system, we must recognize and understand the later changes to the asteroids that result from their own physical properties.

2.b.iii. YORP and Reconfiguration

The Yarkovsky effect is only the first example of a rapidly developing science that connects asteroid shapes, spin states, trajectories, and physical properties to each other. Given the particular importance of shape and spin state, I refer to this field as asteroid morphology.

Yarkovsky connected the trajectory and orbital evolution of an object to the surface thermal properties and to its mass, spin state, and shape. Given a non-symmetric shape, inevitable for a rubble-pile object, shape and orbit also affect the asteroid's spin state. Asymmetric thermal re-radiation leads to a net torque, accelerating or decelerating the asteroid's rotation rate and changing the pole direction (Fig. 2.4). This is termed the Yarkovsky–O'Keefe–Radzievskii–Paddack effect and is universally abbreviated as YORP (Rubincam 2000). Completely despinning a 500-m near-Earth asteroid from breakup rotation takes ~1 Myr on average (Scheeres 2007a), ~1% of their dynamical lifetime.

To complete this set of interactions, an asteroid's orbit and spin state can also determine its shape, given that the object is a rubble pile with limited internal strength. If YORP steadily increases an object's rotation rate, eventually centrifugal acceleration will overcome the local gravity and the internal strength of the rubble pile and the object's shape will reconfigure to accommodate more angular momentum per unit mass. This typically produces a ridge around the equator (e.g. Ostro et al. 2006, Scheeres et al. 2006). Adding still more angular momentum results in material at the equator exceeding breakup velocity and the asteroid will shed mass. This is the presumed origin of the large number of binary near-Earth asteroids, where the primaries are much larger than the secondaries (Margot et al. 2002, Walsh et al. 2008).

Once an asteroid forms a satellite, the primary's shape changes to accommodate the secondary's tidal perturbations and the effects of YORP change (Cuk 2007, Goldreich & Sari 2009, Cuk & Nesvorny 2010). If YORP continues to accelerate the system, angular momentum can be transferred into the orbit of the secondary until it escapes. If the reconfiguration reverses the direction of the radiation torques, the secondary may recombine with the primary. This chaotic progression of reconfiguration should occur many times over the lifetime of each near-Earth asteroid (Scheeres 2007b).

The near-Earth asteroids are reconfigured by more than just YORP. Impacts also disrupt their shapes. Even collisions so small that no material is ejected from the object can dramatically reconfigure a rubble-pile, due to seismic shaking and depending on its initial shape and internal material properties (Asphaug 2008). Finally, tidal stresses during close planetary flybys cause enough stresses on a rubble pile asteroids to resurface them (Binzel et al. 2010). While tides may only move around the surface material on an object, the resulting changes in the color and albedo distribution of the surface are coupled back to the orbit and spin state via Yarkovsky and YORP.

To understand the history of the asteroids and their role in the overall history of the solar system, and to predict the future behavior of individual asteroids (as for the impact hazard), is to understand all of these coupled interactions. This requires knowing the orbits, sizes, shapes, spin states, and, ideally, the density, composition, and material properties of a representative sample of the near-Earth population. Since extreme instances of Yarkovsky migration, YORP spin-up or spin-down, and shape reconfiguration last less than 1% of an asteroid's lifetime, a truly representative sample must include at least several hundred objects.

To determine an asteroid's current orbit merely requires accurate astrometry from conventional optical imaging, and is a necessary precursor to any further observations. Some compositional information is available from optical and infrared spectroscopy. However, to determine size, shape and spin state normally requires *spatial resolution*. Accurate Yarkovsky and YORP predictions for few-hundred meter asteroids require resolution of roughly 10 m or better (Statler 2009).

The finest resolution observations of asteroids are of course from spacecraft: the NEAR mission to Eros and the Hayabusa mission to Itokawa both obtained global maps with resolution finer than 1 m (Zuber et al. 2000, Demura et al. 2006). Spacecraft observations are also likely the only way to obtain detailed information on the internal structure of rubble-pile objects. However, it is prohibitively expensive to send spacecraft to several hundred asteroids. Ground-based astronomical techniques have a decisive advantage. Currently, one of the most powerful and successful techniques for studying asteroids physical properties is radar astronomy, which I describe in the next chapter.

3. Asteroid Radar Astronomy

Radar astronomy is a set of techniques to study target objects by illuminating them with a radio transmission and analyzing the reflected echoes. By coding the transmission, asteroids and planets can be imaged; in some cases with resolution <10 m. I have used radar astronomy to construct models of the shapes and spin states of four near-Earth asteroids and to study the surface properties of the moons of Mars.

Here I describe radar astronomy in detail and discuss my observational results and their meaning. The material in this chapter is drawn from two review articles (Ostro et al. 2002 & Ostro et al. 2007) and from descriptions of the SHAPE radar modeling software (Magri et al. 2007 & Magri 2010), in addition to the Busch et al. papers describing results for individual objects.

3.a. Transmitters and Receivers

The fundamental relationship for radar astronomy is the radar equation for received echo power:

(3.1)
$$P_r = \frac{P_t G_t A_r a \pi r^2}{(4\pi)^2 D_{\text{target}}^4}$$

The received power P_r is proportional to the transmitted power P_t, the transmitter gain G_t, and the receiver's effective collecting area A_r, and the cross sectional area (πr^2) and radar albedo (a) of the target object. Most importantly, it is also inversely proportional to the fourth power of the distance D_{target}. The total power intercepted by the target drops as D_{target}^2, as does the fraction of the reflected power received by the antenna. It is therefore very much preferable to observe objects when they are at their closest to the Earth and the radar.

Currently there are two regularly operating planetary radars. At the Arecibo Observatory in Puerto Rico (Fig. 3.1a), the 2380-MHz (12.6 cm) radar provides a maximum transmit power of 1 MW; at the Goldstone Deep Space Network 70-m station in the Mojave Desert (Fig. 3.1.b), the 8560-MHz (3.5 cm) radar can transmit 450 kW. While Arecibo's 305-m primary is much larger than that of Goldstone, its longer wavelength and lower aperture efficiency mean that the two antennas have roughly the same gain. In consequence, the Arecibo radar is only roughly $(305/70)^2 \approx 20$ times more sensitive than the Goldstone system. Arecibo's higher sensitivity is complemented by Goldstone's ability to slew to any direction on the sky. While Arecibo can see only ~1/3 of the sky (between declinations of 1.35° S and 38.05° N) and can track a given object for no more than ~2.6 hours, due to the fixed dish, Goldstone can view ~80% of the sky (declinations north of 39.55° S) and track objects for as long as they are above the horizon.

The velocity of the target object relative to the Earth and the radar produces a variable Doppler shift, which can be removed by tuning either the transmitter or the receiver, concentrating the echo power into the smallest possible bandwidth. This bandwidth is determined by the target's rotation. A portion of its surface is moving away from the radar relative to the center-of-mass while another is moving towards it. The resulting Doppler shifts give the echo a bandwidth determined by the target's radius, rotation period P, and the sub-radar latitude θ:

(3.2)
$$\Delta v = \frac{2\pi r}{Pc} v \cos(\theta)$$

10

Figure 3.1. (a) Arecibo Observatory, Puerto Rico. The antenna is a 305-m fixed spherical dish. The transmitter and receivers are mounted inside the Gregorian dome on the platform and moved to permit pointing within roughly 20° of zenith.

Viewing an object from its equatorial plane produces the maximum echo bandwidth, while a view directly along the rotation axis gives no Doppler broadening and zero bandwidth.

By matching the bandwidth of the echo, and increasing the integration time t_{int}, the radar obtains the maximum possible signal-to-noise ratio:

$$(3.3) \qquad SNR \propto P_r t_{int}^{1/2} / \Delta v^{1/2} \propto r^{3/2} A_r^2 P_t a P^{1/2} t_{int}^{1/2} / D_{target}^4$$

3.b. Surface properties

A radar echo contains information about the target's surface properties. In particular, the echo polarization provides information on the structure of the near-surface on the scale of the wavelength; while the overall radar albedo is related to the target's dielectric constant and, by inference, to its near-surface bulk density. The polarization ratio and radar albedo can be mapped across an object, providing

11

Figure 3.1: (b), Goldstone Deep Space Network, Mojave Desert, California, 70-m steerable antenna. The transmitter and receivers are mounted in separate feed horns at the Cassegrainian focus.

information about the distribution of surface features (such as for the polar ice deposits on Mercury, Slade et al. 1992, and for lava flows on Mars, Muhleman et al. 1991). Generally, for asteroid radar observations, only the integrated echo polarization ratio and radar albedo are considered, to provide the maximum possible signal-to-noise.

3.b.i. Polarization ratio and surface roughness

Both the Arecibo and Goldstone radars transmit a circularly polarized signal and receive echoes in both senses of circular polarizations. If the asteroid's echo came entirely from single scattering, then it would be entirely in the opposite-sense-as-transmitted circular polarization (OC). A real object has near-surface roughness on the scale of the wavelength and multiple scattering, leading to some echo power coming back in the same-sense-as-transmitted polarization (SC). The target's SC/OC ratio of echo power is a convenient measure of the degree of near-surface structural complexity.

The polarization ratio of asteroid radar targets varies widely, from less than 0.1 to roughly unity: the mean SC/OC of the current NEA radar sample is 0.34, with a root-mean-square dispersion of 0.25 (Benner et al. 2008). Oddly, the NEAs have significantly higher mean SC/OC than the main-belt asteroids (Magri et al. 1999).

Polarization ration is correlated with spectral type and presumably composition (Benner et al. 2008). The S-type asteroid 1992 SK has a polarization ratio of 0.35 ± 0.05, which is fairly typical for its spectral type as well as the population as a whole. The E-type asteroid 1998 WT24 has one of the highest polarization ratios ever recorded for an asteroid: SC/OC = 0.97 ± 0.10 at both 12.6 cm and 3.5 cm wavelengths. The high ratio implies extreme near-surface structural complexity. Out of a sample size of 214, there are only 18 near-Earth asteroids with SC/OC > 0.8 (Benner et al. 2008). These high polarization objects include all five radar-observed near-Earth asteroids known to belong to spectral type E (presumed to have enstatite achondrite compositions, Bus et al. 2002), including WT24, and others of unknown composition (X-class or no spectral data available).

Connecting the polarization ratio to particular surface morphologies is not easy. SC/OC = 0 implies a smooth surface, where the average surface fluctuations on 10-cm scales are much less than a wavelength. SC/OC = 1 implies equal amounts of odd- and even-reflections, but there are many suitable geometries with scattering either between the rock and vacuum (surface scattering) or between composition changes or voids within the material (volume scattering). The latter requires a low-loss medium, such as high-porosity regolith. My subjective physical intuition says that the surface of WT24 is likely dominated by cm-to-m-size rocks that have complex, jagged shapes – similar to lava flows on Earth (e.g. Campbell 2002). A speculative explanation for the high polarization ratio of the E-types is that the presence of cm-scale enstatite crystals gives a preferential size to the components of a rubble-pile fractured by collisions.

3.b.ii. Radar albedo and near-surface bulk density

The radar albedo is defined as the ratio between the echo power received from an object and that received from a hypothetical perfect reflector at the same distance and with equal cross-sectional area that reflects the echo power isotropically. It is possible to have a radar albedo greater than 1, for an object where coherent backscatter sends a large fraction of the incident flux back in the direction it came from (e.g. a flat mirror normal to the line of sight or a right angle reflector cube, Green 1968). Asteroid radar targets are generally diffusely scattering, and therefore have albedos much less than 1.

For a smooth sphere, the OC radar albedo would equal R, the Fresnel normal-incidence reflectance coefficient, and SC/OC would be zero. For a target that has decimeter-scale 'roughness' within a meter or so of the surface, some of the echo power is converted into SC via single scattering from rough surfaces or via multiple scattering. Some fraction of the OC radar albedo then corresponds to R for a hypothetical smooth component of the surface. Generally, one can write R = OC albedo/b for non-spherical rough objects, where the backscatter gain b is greater than 1. R for a homogenous dielectric halfspace with a perfectly smooth boundary is an increasing function of the bulk density ρ, given that the bulk density reflects the dielectric constant. This is a good approximation for rocky and metallic objects like asteroids, but fails for liquid water or nearly pure water ice.

If we assume $b = 1$, we set an upper bound on R, and therefore an upper bound on the bulk density of the smooth component of the near-surface. For objects with low SC/OC, single back-reflections dominate the echo. However, since a small fraction of the OC power is still due to multiple scattering, upper bounds on R are conservative.

Several approximate $\rho(R)$ formulas for real materials have been derived from empirical results, either by using the radar albedos of asteroids visited by spacecraft or by measuring the radar reflectivity of laboratory powders. Magri et al. (2001) used the surface density of 433 Eros as determined from the NEAR spacecraft to calibrate their $\rho(R)$ relationship. Ostro et al. (1985) and Garvin et al. (1985) measured R for powders of various densities and compositions and found a nearly linear or logarithmic dependence on density. Some of the difference between the Ostro et al. and Garvin et al. relationships (Fig. 3.2) may be due to the density range covered by the measurements. Garvin et al. used powders with bulk densities between 1 and 2.3 g/cm^3, while Ostro et al. sampled densities between 1.5 and 3.5 g/cm^3. These relationships are:

(3.2)
$$\rho = (R/R_{Eros})^{1/2}(3.75 \pm 0.1) \text{ g/cm}^3 \text{ (Magri et al., 2001),}$$
$$\rho = 8.33R + (1.08 \pm 0.1) \text{ g/cm}^3 \text{ (Ostro et al., 1985),}$$
$$\rho = \ln((1 + R^{1/2})/(1 - R^{1/2}))(3.2 \pm 0.1) \text{ g/cm}^3 \text{ (Garvin et al., 1985).}$$

These three equations provide reasonable estimates of near-surface bulk densities for low polarization ratios. Given other constraints on composition, radar density estimates can provide an estimate of the porosity of an object's near-surface. A caution: below I will use near-surface bulk density as a proxy for the overall bulk density of various asteroids. This is a reasonable approximation, based on the known densities of Eros, Itokawa, and 1999 KW4 (Ostro et al. 2006), but may not hold for all objects. For example, a solid block covered with regolith more than several wavelengths thick would have higher global than near-surface bulk density.

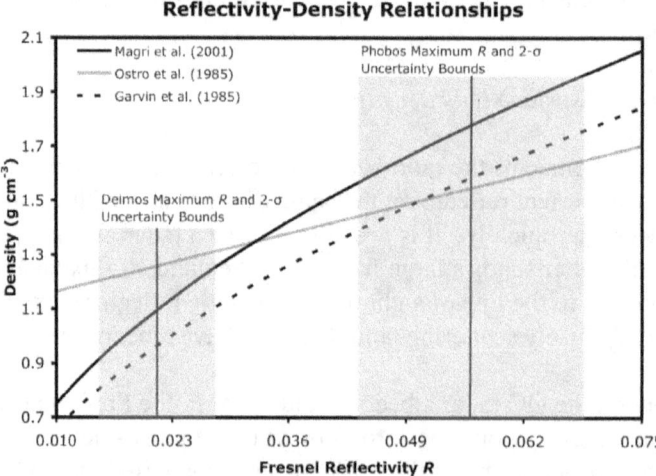

Figure 3.2. Relationships between bulk density d and Fresnel reflection coefficient R, from Garvin et al. (1985), Ostro et al. (1985), and Magri et al. (2001). Upper bounds on R for Phobos and Deimos, with their 2-σ uncertainties, are denoted by gray shading.

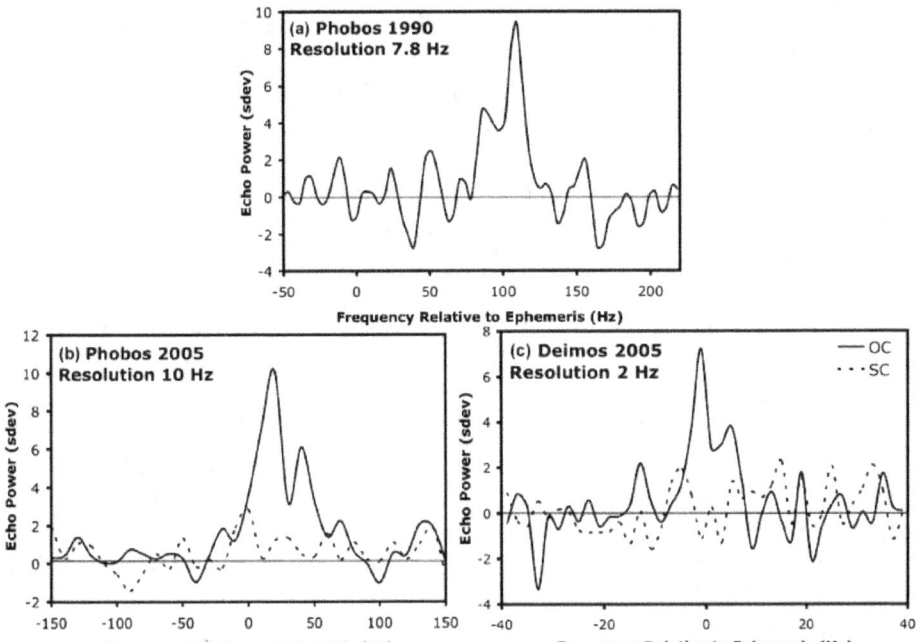

Figure 3.3. Echo power spectra. (a) Phobos from 1990, (b) Phobos from 2005, and (c) Deimos from 2005. The 1990 spectrum is the best single-day detection in that year. The frequency resolution has been smoothed from a raw value of 3.9 to 7.8 Hz in (a), from 5.0 to 10 Hz in (b), and from 0.2 to 2.0 Hz in (c). Echo strength is plotted in standard deviations. The Phobos Doppler prediction ephemerides used an approximate model for the motion of the moon; the observed offsets are not dynamically significant and have no bearing on the albedo and SC/OC results.

If SC/OC is high, then the logic above breaks down: R cannot be estimated because multiple scattering and/or reflections from interfaces that are rough at scales near the wavelength dominate the echo. However, it is possible to constrain the surface density in a relative sense, because the total-power (OC+SC) radar albedo is necessarily strongly related to the near-surface's average bulk density. Thus, for 1998 WT24, the total-power albedo ranks below values obtained for apparently metallic asteroids but above other NEAs, including objects whose global bulk densities have been measured by spacecraft: Eros (2.67 ± 0.03 g/cm^3, Yeomans et al. 2000) and Itokawa (1.9 ± 0.13 g/cm^3, Fujiwara et al. 2006). Therefore, WT24's bulk density is likely between 3 and 5 g/cm^3.

There is no spectral evidence for high iron content minerals on WT24 (Lazzarin et al. 2004) and solid enstatite achondrite has a bulk density of about 3 g/cm^3 (Britt and Consolmagno 2003). All this information suggests that WT24's surface and global bulk densities probably are close to 3 g/cm^3, but multiple-scattering configurations (along with a wide range of possible near-surface porosities) can conspire to give a wide range of total-power radar albedos. For WT24 and similar objects, the radar properties do not strongly constrain either density or porosity.

3.b.iii. Phobos and Deimos

On November 2, 2005, Arecibo observed the moons of Mars, Phobos and Deimos (Fig. 3.3, Busch et al. 2007a). While the moons are both much larger than most near-Earth asteroids (with a mean diameter of 22 km for Phobos and 12 km for Deimos), they were much more distant. Their overall echo strengths were such that I was only able to estimate their disc-integrated radar albedos and polarization ratios. Phobos was previously observed in 1988 and 1990 and the average of the three epochs implies an OC radar albedo of 0.056 ± 0.008 (Ostro et al. 1989). From the 2005 Deimos data, I found an OC radar cross-section of 2.9 ± 0.8 km^2, corresponding to an OC albedo of 0.021 ± 0.006, based on a shape model derived from stereo spacecraft images (Thomas 1999).

SC/OC was 0.17 ± 0.04 for Phobos and 0.12 ± 0.12 for Deimos. These radar properties are unusual when compared to other small bodies (Benner et al. 2008). Phobos' radar albedo is near the low end of the distribution for radar-observed asteroids, while Deimos has the lowest radar albedo of any radar-detected solar system object. Since SC/OC measures near-surface structural complexity on the scale of the radar wavelength, I conclude that the moons' surfaces are low density and relatively smooth on centimeter to meter scales.

In the albedo regime of Phobos and Deimos, the density-reflectivity relationships in Eq. (3.2) are nearly linear (Fig. 3.2). The Garvin et al. formula gives a near-surface density of 0.9 ± 0.2 g/cm^3 for Deimos, the Ostro et al. formula gives 1.2 ± 0.2 g/cm^3, and the Magri et al. formula gives 1.1 ± 0.2 g/cm^3. There are similar differences for Phobos. In light of all the available information, I estimate an upper bound on Deimos' near-surface bulk density of 1.1 ± 0.3 g/cm^3 and an upper bound on Phobos' near-surface bulk density of 1.6 ± 0.3 g/cm^3. In both cases, the near-surface density appears to be less than the bulk density of the moon.

Phobos and Deimos have very similar optical and infrared spectra and inferred surface compositions, although Phobos has greater structural diversity (Thomas et al. 1999). Laboratory spectral analogs include lunar soils and heated, dehydrated, carbonaceous chondrites (Rivkin et al. 2002), both of which have grain densities about 2.7 g/cm^3 (Britt & Consolmagno 2003). For an assumed grain density of 2.7 g/cm^3, the surface bulk density constraints imply mean near-surface porosities of at least (40 ± 10)% on Phobos and (60 ± 10)% on Deimos (porosity = 1 − bulk density/grain density). Neither porosity is implausible: for grains tens of microns in size, porosities near 70% are possible, either from electrostatic repulsion (e.g., Gold 1962), or because of grain shape effects (e.g., very angular particles; Latham et al. 2002). Still, why does Deimos have a higher surface porosity than Phobos, and why are the surface densities of Phobos and Deimos so low when compared to other radar targets?

Phobos and Deimos are in a unique dynamical environment: impact ejecta reaching escape velocity go into Mars orbit rather than escaping completely (Thomas et al. 1986; Veverka et al. 1986). This impact debris can then form very diffuse (100s of particles/km^3) dust bands around Mars (Soter 1971; Krivov and Hamilton 1997). Particles smaller than about 30 μm are not stable on timescales of 100 yr in Phobos-like orbits: solar radiation pressure increases orbital eccentricities until the particles encounter Mars' atmosphere (Hamilton and Krivov 1996). Similarly, particles smaller than ~15 μm cannot remain in Deimos-like orbits around Mars. However, larger particles remain in well-defined bands for longer durations and can be re-accreted (Krivov and Hamilton 1997). The velocity of dust-band particles during accretion is very low, comparable to the escape velocity, which is less than 20 m/s for Phobos (Davis et al. 1981; Veverka et al. 1986). Such low accretion velocities may lead to the formation of high-porosity regolith, because particles settling relatively gently are less likely to compress the material they land on.

Differences in regolith particle size can also contribute to the satellites' different near-surface densities. Phobos may have few particles smaller than 30 μm, while Deimos can have particles down to

16

~15 µm. 20 µm grains are small enough for electrostatic repulsion to possibly become significant enough to increase regolith porosity. Also, Deimos re-accretes any dust grains more efficiently than Phobos. Phobos' Hill radius (the approximate distance where tidal forces from Mars prevent any re-accretion) is 17 km, while Deimos' is 26 km (Deimos' lower mass is more than compensated by its greater distance from Mars). Deimos' 50% lower surface gravity may also contribute to higher regolith porosity. This combination of accretion volume and velocity, particle size, and surface gravity can plausibly produce the low bulk density inferred from the Arecibo data. In contrast, small near-Earth asteroids do not retain the majority of their ejecta (Asphaug & Nolan 1992) and only acquire material in high-velocity collisions.

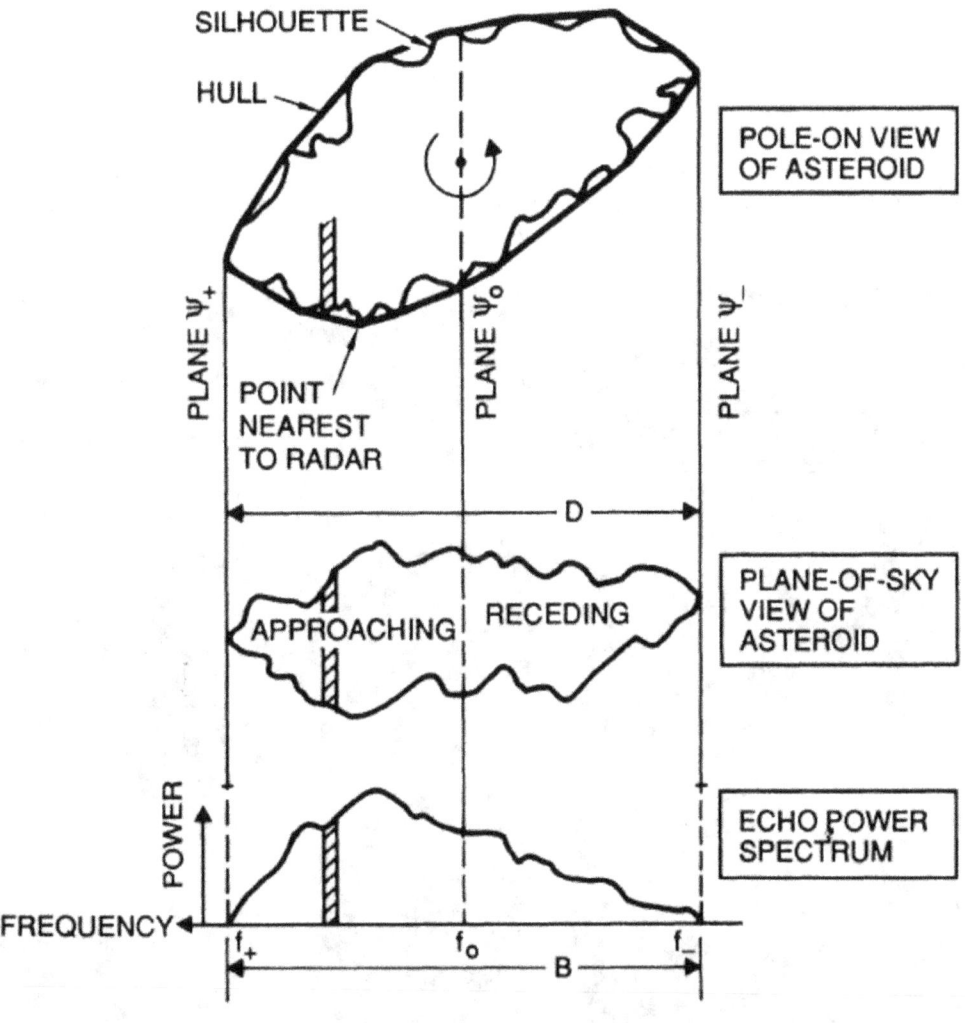

Figure 3.4: Schematic of Doppler-only radar resolution. A monochromatic radar beam is transmitted from the ground at a specified carrier frequency, reflected from the target object, and separated out in frequency. The Doppler shift of the center-of-mass is determined by the object's geocentric trajectory, the position of the receive station on the Earth, and transmitter tuning, while the Doppler width of the echo (exaggerated in the plot) is determined by the target's size, rotation rate, and pole direction. Picture from Ostro 1993.

3.c. Doppler resolution and range coding

The Doppler broadening of the radar echo resolves the target in frequency, corresponding to line-of-sight velocity (Fig. 3.4, Ostro et al. 2002). The radar echo can be resolved in time delay, corresponding to line-of-sight distance, by coding the transmitted signal (e.g. Cohen 1991), and in Doppler shift (Ostro et al. 2002).

Combining Doppler and delay resolution yields two-dimensional delay-Doppler images (Fig. 3.5), the principal data product of radar astronomy. Currently, the Arecibo radar transmitter can be coded at a maximum range resolution of 7.5 m, while the Goldstone radar can be coded to provide range resolution of 3.75 m (Slade et al. 2009).

The resolution of a delay-Doppler image is normally quoted in time delay and frequency rather than distance and velocity, converting from one set of units to the other using the speed of light and the carrier frequency. A resolution of 7.5 m x 1 cm/s at 2380 MHz, for example, is equivalent to 0.05 μs x 0.0794 Hz. Note that the resolution in range is (time-delay resolution) * c/2. The factor of 2 comes from the radar beam traversing the path from transmitter-to-object twice, once before and once after being reflected.

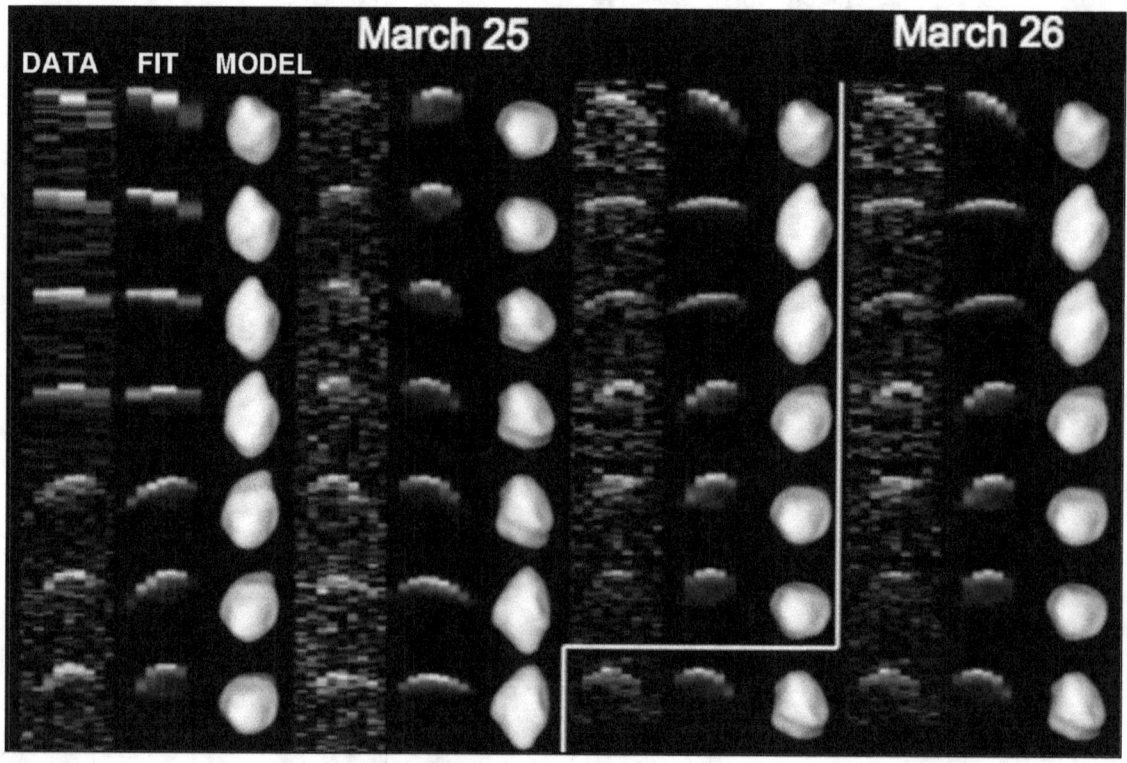

Figure 3.5: Delay-Doppler images of (10115) 1992 SK from March 25 & 26 1999, with fits and plane-of-sky projections of the model. Within each delay-Doppler image, time delay of the echo increases from top to bottom, with a scale of 0.25 μs/pixel (37.5 m/pixel). Doppler frequency increases from left to right, with resolution 5 Hz for the first 4 images on March 25 and 2 Hz for all other frames. The model projections have a plane-of-sky width of 1.5 km and are oriented with north upward and east leftward.

For single-station ("monostatic") observations, the radar must cycle between transmit and receive. The transmitter runs for slightly less than the round-trip light-travel time for the radar beam to reach the target, be reflected, and return to Earth. Then the system switches to receive for the same amount of time. This limits the frequency resolution of the resulting delay-Doppler images to one over the round-trip time less the changeover. For Goldstone, switching from transmit to receive takes a minimum of 3 s. Thus a target at 0.01 AU, with round-trip time of 10 s, could only be imaged with frequency resolution of $1/(10\ s - 3\ s) = 0.143$ Hz.

For very close targets, the radars are operated in "bistatic" mode, with transmit from either Arecibo or Goldstone and receive with any of a number of radio telescopes. This relaxes the minimum-bandwidth restriction, and is in a sense the prototype for the multiple-station radar techniques I have developed (Chap. 4 & 5).

3.d. Shape modeling

To obtain a delay-Doppler image, the target's trajectory must already be known precisely from optical astrometry. Since the radars have very narrow beams (roughly 1 arcminute), they are not useful tools for discovering NEAs. If the uncertainty in an object's plane-of-sky position at any given time is much less than the Arecibo or Goldstone beamwidth, it can be detected (typical optical astrometry provides pointing to ~1 arc*second*). Given that its line-of-sight velocity is known well enough for it to be tracked without blurring in range and/or Doppler shift, it can be imaged. In practice, radar observing campaigns consist of iterations of observation and trajectory refinement based on radar astrometry at increasing Doppler and delay resolution.

An individual delay-Doppler image provides considerable information. Delay-Doppler astrometry is the most accurate technique to determine an object's position in space, other than spacecraft telemetry. While any individual measurement, i.e., from one image, only collapses the uncertainty region in two dimensions (line-of-sight distance and line-of-sight-velocity), multiple epochs of astrometry provide measurements at different angles. For comparison: without adaptive optics, optical astrometry is uncertain by 0.1"-1". A delay measurement uncertainty of 7.5 m for an object at 0.01 AU is equivalent to 0.001". For application to the asteroid impact hazard, radar astrometry on average increases the interval for which potential impacts can be predicted by about 300 years for recently discovered objects (Ostro & Giorgini 2004, Giorgini et al. 2009).

A single image also provides information on the target's physical properties: the echo's range extent places a lower bound on its size, and given a rotation period from optical lightcurves and a size estimate, the echo bandwidth corresponds to the subradar latitude (the angle between the radar-asteroid line and the asteroid's equatorial plane). In even low-resolution images, the asteroid's overall shape and large surface features are evident. For example, Fig. 3.5 clearly shows that the asteroid 1992 SK is elongated and asymmetric. Combined with the target's size, the total echo strength provides an estimate of the radar albedo, which constrains the near-surface bulk density as described above.

To produce a complete and unique estimate of a target's shape and spin state requires multiple images from different orientations. Different orientations are required both to insure that as much of the object as possible is seen and because any given delay-Doppler image is a non-unique mapping of the surface. For a sphere viewed from the equator, each point in the northern hemisphere plots to the same point in delay-Doppler space as a point in the southern. More complicated shapes can have regions of three- or more-to-one projection. Given a series of images at a range of subradar latitudes and longitudes, the different delay-Doppler trajectories of each point on the object can be disentangled and a unique model of the asteroid's shape and spin state constructed (Ostro et al. 2002).

Currently, delay-Doppler shape inversion uses a software package called SHAPE, which semi-autonomously fits shape models to radar data.

3.d.i. SHAPE *software*

SHAPE, originally written by R. Scott Hudson (Hudson 1993) and more recently updated and documented by Christopher Magri (Magri et al. 2007, Magri 2010), uses constrained weighted least squares to estimate parameters describing the shape of an object, its rotation state, its radar- and optical-scattering properties, and corrections to the delay-Doppler trajectory prediction. The input to SHAPE is a series of delay-Doppler images, Doppler-only echo spectra (as in Fig. 3.2), optical photometric measurements, and initial model parameters.

An asteroid shape model typically contains many tens to thousands of parameters: the shape can be represented as an ellipsoid (3 parameters), a sum of spherical harmonics (generally a few dozen to 100), or as a polyhedron with many vertices (several hundred to over 10000). For each adjustment to a single parameter, SHAPE must compute the model's orientation relative to the radar at each observing epoch; generate synthetic radar data for that geometry (a fit); compute the chi-square of the fit to the delay-Doppler image, echo spectrum, or lightcurve point; and add the chi-squares for all observations together. An overall grid search is computationally impossible, and currently SHAPE adjusts parameters in sequence: e.g. fitting the position of one vertex in a polyhedral representation of the asteroid's shape, followed by the next vertex, until it runs out of model parameters. Then it repeats the process until chi-square has converged to a user-specified level, locating a local minimum of goodness-of-fit in the model parameter space.

The sequential adjustment of parameters produces artifacts in the fits, such as linear features of vertices that were adjusted one after the other (Magri et al. 2007, Bramson et al. 2009) or 'sea urchin' models where individual facets are moved to non-physical locations, giving a shape covered in large spikes that coincidentally fits the few brightest delay-Doppler noise pixels. To steer models away from non-physical shapes, SHAPE includes data weighting and a series of penalty functions (Magri et al. 2007).

Data weighting is straightforward. If there is, for example, a series of high-resolution Arecibo images that show a surface feature such as a crater, the user can specify that those images be weighted higher in the chi-square sum. Then SHAPE will fit the surface feature, perhaps at the expense of matching lower-weighted data.

Penalty functions are somewhat subtle. To prevent sea urchin models, SHAPE uses the nonsmooth penalty. For each pair of adjacent facets, SHAPE computes the angle θ between the surface normal vectors. It then averages $(1 - \cos(\theta))^4$ over the entire shape, multiplies that average by a user-specified penalty weight, and adds the entire number to the chi-square. Using nonsmooth suppresses facet-scale structure: a perfectly smooth sphere or a flat planar object would both be close to the minimum possible nonsmooth value. Other penalty functions included in SHAPE suppress concavities, oblateness, the offset between the model's center-of-figure and the object's center-of-mass, and the model's moment-of-inertia tensor being non-diagonal.

In addition to suppressing fitting artifacts, the penalty functions determine how SHAPE fills in the shape of the object in areas that did not provide strong echoes (either not visible or observed only at grazing incidence). These regions are highlighted in SHAPE's output, and must not occupy a large fraction of the surface area or volume of the object for the overall shape of the object (and hence Yarkovsky and YORP modeling) to be reliable.

Changing the penalty weights will change the outcome of a SHAPE run with the same input data and data weighting; as will changing the starting model's pole direction, rotation rate, or shape. In consequence, fitting delay-Doppler data with SHAPE generally consists of an initial grid search with an ellipsoid or coarse-resolution harmonic shape model to determine potential pole directions from the echo's overall bandwidth and range extent; followed by high-resolution polyhedron shape models with a variety of different weights on both the data and penalty functions.

As an illustration of the importance of the choice of the penalty weights, data weighting, and starting model, see Figs. 3.6 and 3.7. They show three different shape models of the asteroid 2008 EV5, with pole directions determined from the echo bandwidth as a function of time and then allowed to vary over a 5° range to reflect the uncertainty in the bandwidth measurements. While all three models have a ridge within 30° of the equator and a concavity breaking the line of that ridge, the three shapes are very different. All provide fits to the delay-Doppler data that have approximately equal reduced chi-square and are formally equally good estimates of EV5's shape.

To a fairly significant extent, the choice of a preferred shape is subjective. While good fits to the EV5 images require a ridge and a 150 m concavity, the ridge can be of widely varying height and centered on any latitude within 30° of the equator. I selected the nominal shape to minimize the overall oblateness (effectively, the magnitude of the ridge) while also minimizing the facet-scale structure (via the nonsmooth penalty described above). Reducing the penalty for oblateness and starting with a model with a modest equatorial bulge produces a ridge so large that the final model resembles a flying saucer. A higher penalty for oblateness and lower penalty for facet-to-facet structure almost removes the ridge and produces a random pattern of small lumpy features.

The nominal shape is a conservative estimate of EV5's topography. While the real surface must have 10-m scale structures, SHAPE will not produce an accurate model of them. I therefore chose to not impose any such topography onto the model. Likewise, while EV5's ridge is most likely more irregular than in the nominal model, I chose to avoid geologically radical shapes unless *required* to fit the data, with the risk of suppressing real structure. This caution applies to all of the objects I have modeled with SHAPE, particularly to estimates of Yarkovsky accelerations and YORP torques.

To date, SHAPE has been used to construct shape models of 30 asteroids (Benner 2010). The four near-Earth asteroids I personally have studied with SHAPE are 1992 SK, 1998 WT24, 1950 DA, and 2008 EV5.

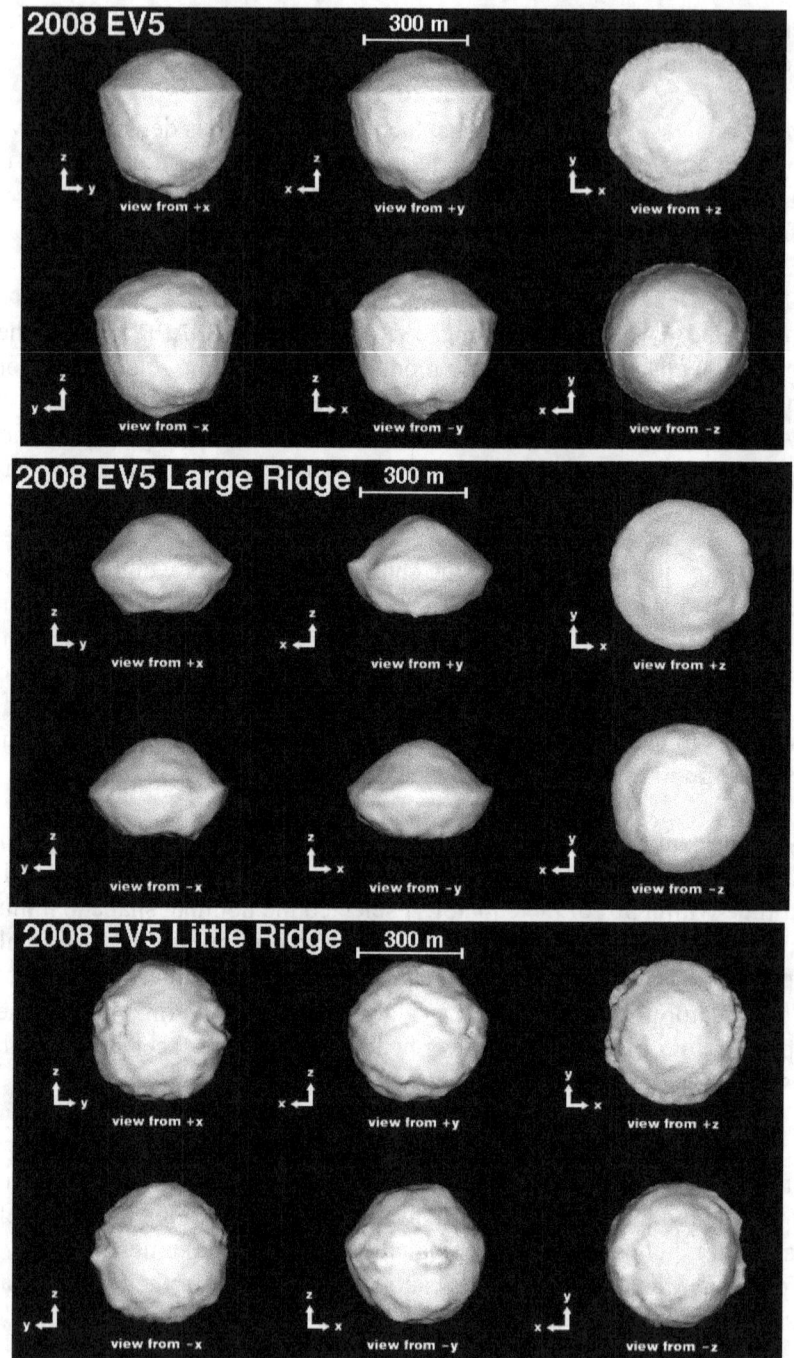

Figure 3.6: Model of 2008 EV5 (top) and two alternate shapes, one with a large ridge (middle) and one with almost no ridge (bottom). These illustrate the range of potential shapes that are equally good matches to the data (Fig. 3.10). Each model is viewed from six orthogonal directions, along its principal axes. Rotation is around the z-axis, with +z in the direction of the angular momentum vector. Yellow-shaded regions were either seen at incidence angles >60° or not seen at all.

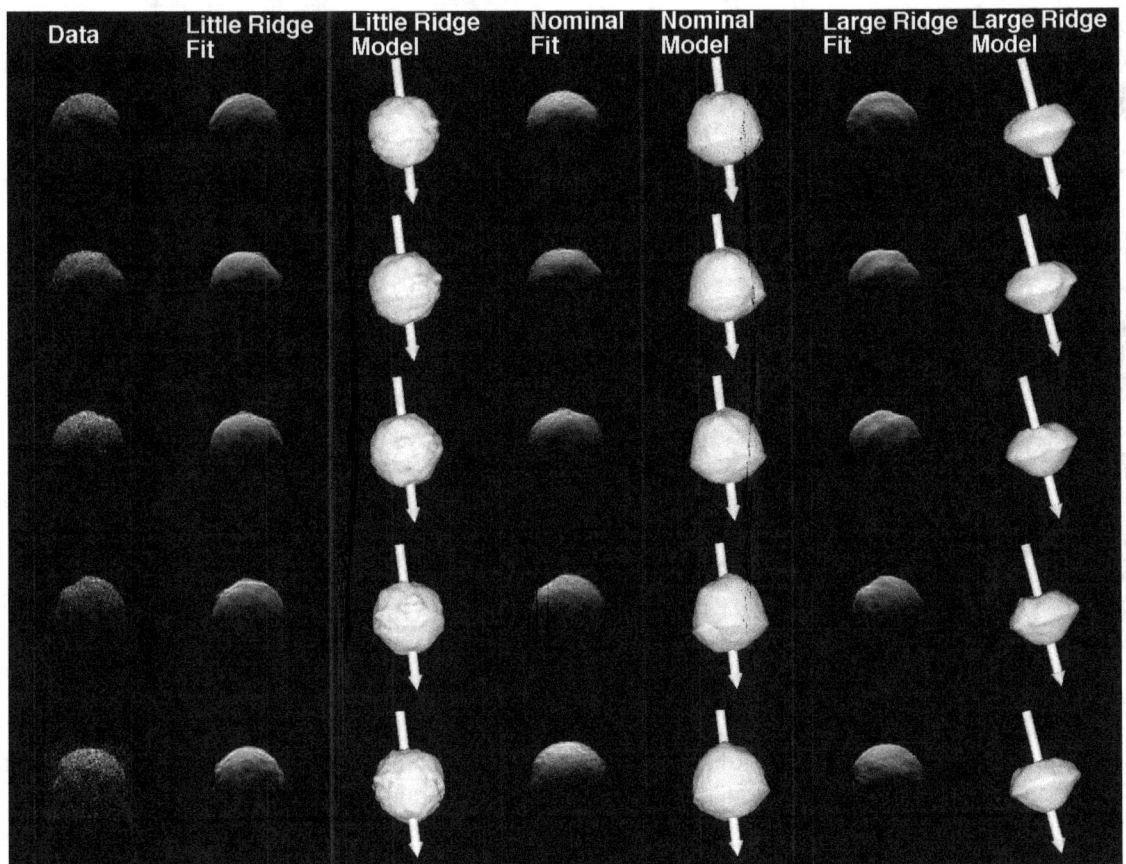

Figure 3.7: Selected delay-Doppler images of 2008 EV5, and SHAPE *fits and plane-of-sky views for the three models in Fig. 3.4. The arrows are the models' spin axes, and are not included in the shape or the fit images. These images span slightly over half of a rotation. Images are oriented as in Fig. 3.3.*

3.d.ii. 1992 SK

10115 (1992 SK) is an S-class near-Earth asteroid (Binzel et al. 2004) that was discovered by E. Helin and J. Alu at Palomar in September 1992 (Helin 1992). Subsequent optical observations and investigation of archival plates, including a precovery from 1952, refined the orbit and indicated that SK would make an approach to within 0.06 AU in 1999. Goldstone observations during 1999 March 22-27 provided the delay-Doppler images shown in Fig. 3.3, which I used to construct the shape model (Busch et al. 2006).

My pole direction estimate for SK implies that the observations took place at sub-radar latitudes between -20° and -40°. Therefore, the shape of the asteroid south of about 60° north latitude is well constrained (Fig. 3.8). The regions shaded in yellow in Fig. 3.8 are those that are relatively poorly constrained: they were either seen at incidence angles greater than 60°, so that the echo from them was relatively weak, or were never seen at all. The north polar concavity, for example, may be filled in without affecting the quality of the fits and would increase the model's volume by ~3%.

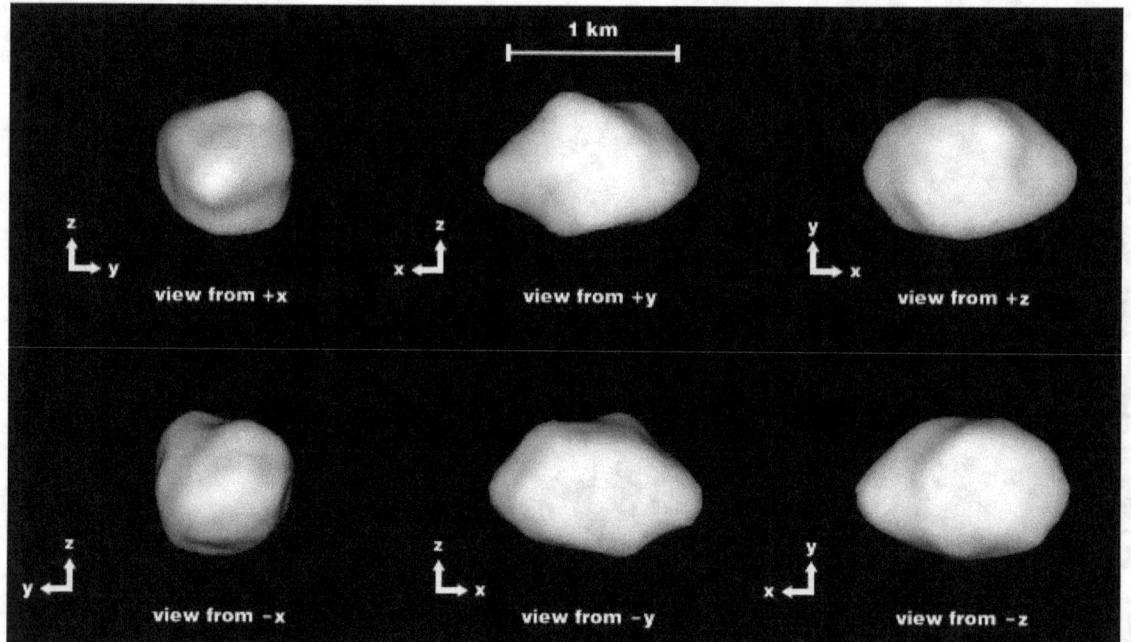

Figure 3.8. Principal axis projections of the 1992 SK model. Much of the northern half of the object was not seen. This model contains 510 vertices. Figure from Busch et al. 2006.

SK illustrates the limitations of delay-Doppler data in obtaining pole directions: using the radar data only, there are five widely separated pole directions that produce comparably good fits. Fortunately, P. Pravec at Ondřejov Observatory in the Czech Republic obtained R-band optical lightcurves as part of his NEA photometry program (Pravec et al. 1998) during February and March 1999. The lightcurves indicate a rotation period of 7.3182 ± 0.0003 hours and suggest an elongated, mildly asymmetric shape. The lightcurves have amplitude 0.6-1.0 mag. Three of the five pole directions do not produce the required lightcurve amplitude and therefore can be rejected. The remaining two solutions are separated by ~180° on the sky. Including the lightcurves in the fitting splits the ambiguity between the mirror solutions and locates the pole direction to within 5° of ecliptic longitude, latitude = (99°, -3°). Table 3.1 lists some properties of the SK shape model (Busch et al. 2006).

Using methods described in Scheeres et al. (1996), I studied the dynamical properties of the SK model. Assuming that the shape model has a uniform internal density, the surface acceleration (gravity + rotational acceleration), escape velocity, and gravitational slope can be mapped across the asteroid (gravitational slope is defined as the angle between the local acceleration vector and surface normal). Such maps provide insights into the asteroid's surface features and likely history (Fig. 3.9).

For SK, the nominal model has a bulk density of 2.3 g/cm^3 (based on the near-surface density inferred from the radar albedo, Sec. 3.d). For this density, SK's surface is very subdued, with a maximum slope of 31° and an average slope of 11°. The maximum slope is less than the angle of repose for granular material - the model does not constrain the subsurface structure of the asteroid. The surface acceleration varies over (2.30 to 3.26) x 10^{-4} m/s^2. The escape speed from the asteroid varies from 0.26 to 0.37 m/s across the surface. To the roughly 150 m resolution of the model, SK is the classic vision of an asteroid: a gravity-dominated pile of rocks.

Table 3.1. 1992 SK Model Properties

OC Radar Albedo: 0.13 ± 25%
Equivalent Spherical Radar Albedo: 0.11 ± 0.02
Near-Surface Bulk Density: <= 2.3 g/cm^3
Spin State
    Pole Direction, J2000 ecliptic: (99°±5°, -3°±5°)
    Sidereal Period: (7.3182 ± 0.0003) h
Volume: 0.53 km^3 ± 40%
Model Surface Area: 3.4 km^2 ± 30%
Diameter of equal-volume sphere: 1.0 km ± 20%
Maximum Dimensions:
    x-axis: 1.39 km ± 20%
    y-axis: 0.90 km ± 20%
    z-axis: 0.91 km ± 20%

Properties of the 1992 SK shape model. The radar albedo measurements are typical for an S-type silicate asteroid and imply a near-surface bulk density of ~2.3 g/cm^3 (see Sec. 3.d). From Busch et al. 2006.

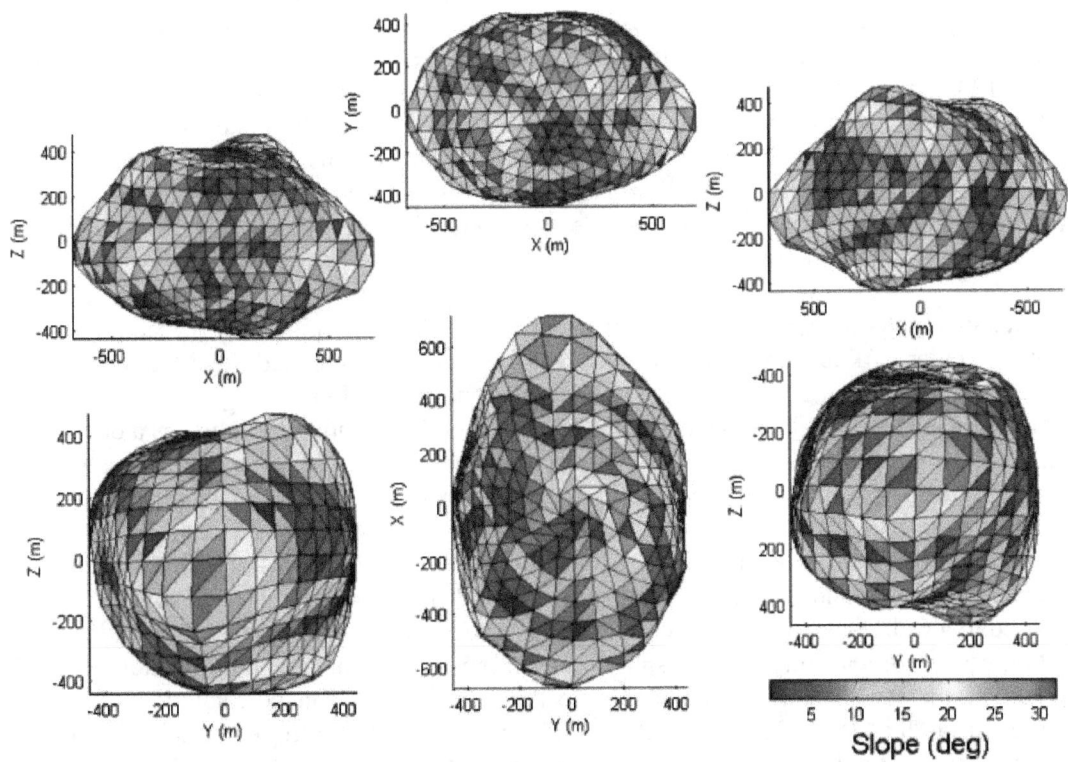

Figure 3.9. Principal axis views of the 1992 SK shape model, showing gravitational slope mapped over the surface, assuming a bulk density of 2.3 g/cm^3. Figure from Busch et al. 2006.

Given SK's shape, pole direction, and bulk density, and estimates of the thermal properties of its near-surface, the methods of Scheeres 2007a and Scheeres & Mirrahimi 2008 can be applied to estimate the magnitude of the YORP torque. SK's rotation rate is currently decreasing. For a density of 2.3 g/cm^3 and nominal thermal properties, it is slowing down at a rate of ~1.5e-10 rad/s/year; such that it will be despun completely and begin rotating in a different direction in between 1 and 2 million years. However, the YORP prediction is very dependent on SK's shape at scales smaller than our model's resolution (Statler 2009). The true YORP spin-down of SK can potentially be measured as early as the late 2010's, when the change in period will have accumulated to about one rotation from the 1999 observations. Similar measurements have already been reported for the aptly-named asteroid 54509 YORP (Taylor et al. 2007).

3.d.iii. 1998 WT24

The Earth/Venus/Mercury-crossing asteroid (33342) 1998 WT24 was discovered by LINEAR on 1998 December 4. It has semi-major axis 0.718 AU, eccentricity 0.418, inclination 7.34° and is near a 2:5 resonance with Mercury, a 1:1 resonance with Venus, and a 5:3 resonance with Earth. WT24 currently makes close approaches to all three planets. It also has a very low flyby delta-v (the velocity change from low Earth orbit to a flyby trajectory), potentially making it an appealing target for examination by a spacecraft, although the delta-v for rendezvous is high.

During its close Earth approach in December 2001, WT24 was observed extensively using thermal infrared radiometry (Harris et al. 2007), optical photometry and spectroscopy, and radar (Busch et al. 2008). Optical observations gave the rotation period, 3.697 ± 0.001 h (Krugly et al. 2002), and showed that WT24 is of spectral class E (Lazzarin et al. 2004). During the approach, within 0.0125 AU (4.9 lunar distances), WT24's radar echo had SNR measured in the tens of thousands for a single transmit-receive cycle. The radar observations used many different stations: delay-Doppler imaging using Goldstone (both monostatic with the 70-m DSS 14 and bistatic between DSS 14 and the 34-m DSS 13) and Arecibo, as well as continuous-wave Doppler-resolved echoes received at several additional stations (Di Martino et al. 2004).

The delay-Doppler images of WT24 show considerable structure (Fig. 3.10, Appendix 1.a). In particular, a large radar-dark feature interpreted as a concavity and an associated ridge-like feature are visible at one set of rotation phases and a small but conspicuous group of radar-bright pixels, which I interpret as a raised feature, is evident beyond the leading edge of the echo. The echo has a delay depth of ~1.5μs (225 m), suggesting a diameter of a few hundred meters. The echo bandwidth is roughly constant (varying by no more than 15%), implying that the asteroid is not very elongated.

The early stages of my modeling included a grid search over the entire sky with a resolution of 15°, stepping down to 3° around candidate poles. The asteroid's motion across 110° of sky during the observations led to a well-constrained pole direction; it was seen from a wide range of sub-radar latitudes. Progressing from an ellipsoid to a spherical harmonic representation and ultimately to a 4000-vertex polyhedron, I arrived at a unique model that fits the radar images. The pole solution, (15°, -22°) ± 5°, differs by 34° from the (355°, -52°) ± 30° inferred by Harris et al. (2007) from thermal infrared lightcurve observations. I did not fit the model to lightcurve data provided by Krugly et al. (2002) or Pravec (2010), because those data were not presented with the times of individual lightcurve points, but the shape model qualitatively matches the structure of the lightcurve with an optical scattering law that contains minimal limb darkening.

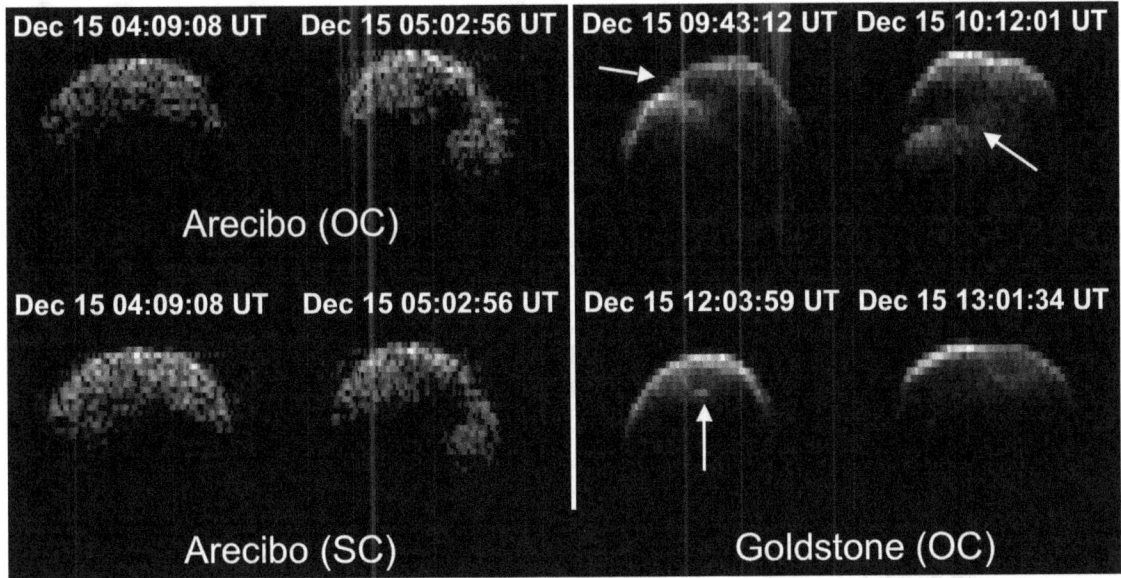

Figure 3.10: Selected delay-Doppler images of 1998 WT24. The arrows indicate the radar-dark feature interpreted as a concavity, an associated ridge-like feature, and a raised feature beyond the leading edge of the echo. For the full set of radar images and corresponding fits, see Appendix 1.a. Figure from Busch et al. 2008.

There are some subtle structures in the images that the model does not fit, particularly near the low- and high-Doppler limits of the echo. At some level, SHAPE's model of radar scattering is simplistic: one scattering law will not hold over the entire surface and a cosine diffuse scattering law is not exactly correct everywhere. This makes fitting the extremes of the echo problematic, and may be reflected in facet-scale errors in the model. The model's principal-axis extents are (470 x 425 x 400) ± 40 m (Table 3.2), where the uncertainty is dominated by the Doppler resolution. The shape is dominated by three concavities, comparable in size to the extent of the object (Fig. 3.11), reminiscent of the impact craters seen on 433 Eros (Veverka et al. 2000) and 253 Mathilde (Veverka et al. 1999) although much smaller in scale.

Given the model's dimensions and the estimates of the asteroid's radar cross section (Table 3.2) and absolute magnitude H, I can estimate WT24's radar and optical albedos, where radar albedo = radar cross section/cross-sectional area and optical albedo = $10^{-0.4H}$(1329 km/effective diameter)^2 (Pravec and Harris 2007). WT24's radar albedo, 0.42, is towards the high end of the distribution of asteroid radar targets and implies a near-surface density of about 3 g/cm^3 (Sec. 3.b, Benner et al. 2008). Taking H = 18.69 ± 0.3 (Kiselev et al. 2002), WT24 has a very high optical albedo for an asteroid: 0.34 ± 0.2 (for comparison, Harris et al. 2007 obtained 0.56 ± 0.2), corresponding to the reflectance of the enstatite achondrites and no other major meteorite class (Lazzarin et al. 2004). Therefore, WT24 is most likely of enstatite achondrite composition.

WT24's radar echo has approximately equal power in both circular polarizations, indicating an extremely rough surface on centimeter to meter scales (see Sec. 3.b). Despite its extreme roughness at radar wavelengths, on scales of 10 m and larger the surface of WT24 is relatively subdued (Fig. 3.11). Assuming a density of 3 g/cm^3, the average facet-scale gravitational slope (angle between the inward surface normal and the local acceleration vector) is 12°, and the maximum slope is 40°. The equator and

Table 3.2: 1998 WT24 Model Properties

Principal-axis dimensions: (470 x 425 x 400) ± 40 m
DEEVE dimensions: (454 x 417 x 378) ± 40 m
Equivalent diameter: 415 ± 40 m
Surface Area: 0.57 ± 0.12 km^2
Volume: 0.038 ± 0.01 km^3

Pole direction, J2000 ecliptic: (15º, -22º) ± 5º
Rotation period: 3.6970 ± 0.0002 h

Mean gravitational slope: 11.6º
Maximum gravitational slope: 40º
Range in potential across surface: 0.14 m/s.

Mean radar cross-section: 0.027 ± 0.003 km^2 (S-band, 13-cm)
Total power (SC+OC) radar albedo: 0.42 ± 0.04

Optical albedo: 0.34 ± 0.20

Properties of the 1998 WT24 shape model. DEEVE = dynamically-equivalent equal-volume ellipsoid, an ellipsoid with the same volume and moments of inertia as the shape model. Gravitational slope and potential calculations assume a bulk density of 3 g/cm^3.

the depression near the south pole are the lowest potential regions. While the necessary escape speed (for which a particle launched away from the surface at a given location on any trajectory is guaranteed to escape) varies between 19 and 33 cm/s, the sufficient escape speed (at which there is some escape trajectory from a given location) is between 0 and 13 cm/s. The zero in sufficient escape speed corresponds to the rotational Roche lobe intersecting the asteroid's surface near -70° latitude and 30° longitude, near the highest point of the potential. In this area, an object disturbed with minimal speed can in principle escape from the asteroid. It should be free of fine-grained regolith.

Assuming that WT24 remains in a principal-axis rotation state, as it is now, and that the asteroid's thermal properties are constant over its surface, and assuming that the asteroid's orbit is constant, WT24's YORP torque can be estimated. Due to rapid evolution of the orbit due to the regular planetary encounters, any such estimate is only accurate for at most the last million years (NEODys 2010). For the thermal properties measured by Harris et al. (2007), WT24 is currently spinning down at an average rate of 2×10^{-7} deg/s/yr, again with the caveat of the resolution of the shape model. At this rate the asteroid will be despun in approximately 150 kyr. Extrapolating backwards in time at this rate, 75 kyr ago the asteroid had a period of 2.5 hours, which would place some regions of its surface at orbital velocity, suggesting that if the asteroid is composed of several monolithic components, these may have been in orbit about each other. If so, WT24 went through a period of reconfiguration.

However, simply extrapolating the current deceleration rate forwards and backwards in time is not valid, because the obliquity of the asteroid is also changing, and hence so are the YORP torques. The obliquity changes are sensitive to the asteroid's thermal properties, in particular the average thermal

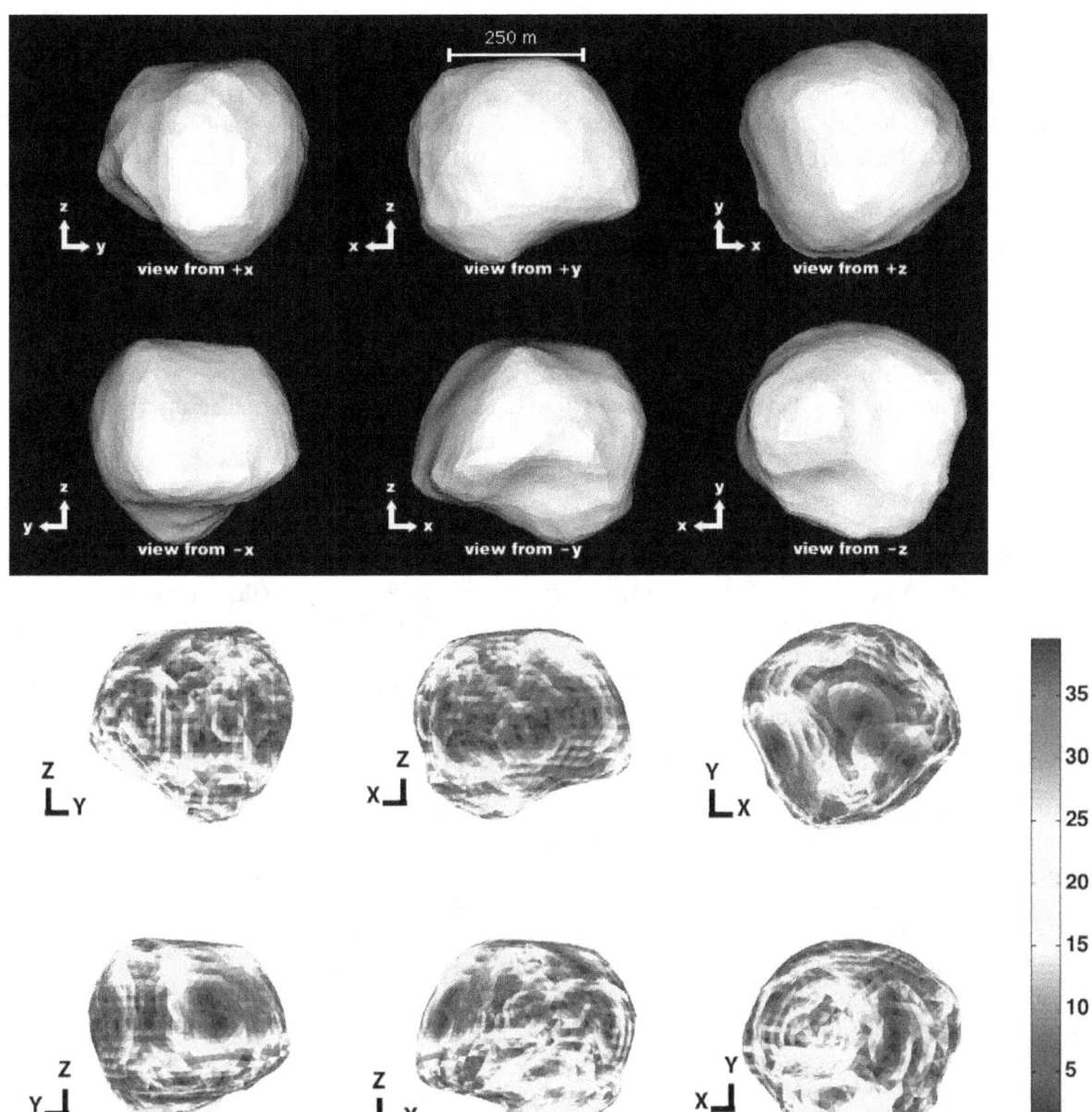

Figure 3.11. 1998 WT24 model in principal axis projection (top) and shaded for gravitational slope assuming a uniform bulk density of 3 g/cm³ (bottom). For lower densities, the peaks near the north and south poles are outside the energetic Roche lobe (surface acceleration becomes negative), so loose material in these areas may escape the surface and go directly into orbit. Slope is given in degrees relative to the outward surface normal.

lag angle (the angle between the sub-solar point and the highest temperature location). Using the Harris et al. (2007) estimate of WT24's thermal inertia and the Rubincam (1995) definition of the thermal lagangle in terms of thermal inertia and average equilibrium temperature, the average thermal lag angle of WT24 is between 1° and 10°. If the average thermal lag angle exceeds ~6°, in the recent past WT24 was spinning at breakup with obliquity near 90°, and will continue to spin down. If the lag angle is low, it was spinning *slower* in the past, with obliquity ~180°, and will spin up in the future. An intermediate lag angle (close to 6°) can balance between these two cases. To determine the current thermal lag angle more precisely would require a thermal model incorporating the detailed shape of the object, and would illuminate the spin-state history of the object. I am currently considering the possibility of using (sparsely sampled) thermal lightcurves of WT24 obtained from the WISE spacecraft for this purpose.

If WT24 was once spinning at its breakup rate, the internal structure of the asteroid may be several pieces resting on each other that were previously in mutual orbit. In this case, the asteroid's surface is too young to have three large craters, and the basins are either preexisting on the fragments or simply reflect the shape of the reaccumulated blocks resting on each other.

As with 1992 SK, the YORP acceleration of WT24 should be detectable in the next several years. In particular, WT24 will next be close enough to Earth for high-resolution radar observations in 2015.

3.d.iv. 1950 DA

(29075) 1950 DA was discovered on February 23, 1950 (Wirtanen 1950), observed for 17 days, and then lost until December 21, 2000, when an object was discovered and given the provisional designation 2000 YK66, but then recognized as being 1950 DA (LONEOS 2001, Bardwell 2001). It was the subject of extensive optical photometric (Pravec et al. 1998, Pravec 2010) and Goldstone and Arecibo radar observations (Giorgini et al. 2002) before and during a 0.05 AU approach to the Earth on March 5, 2001. While most of the interest in 1950 DA is due to its trajectory (Sec. 3.e.i), it also has an anomalously high radar albedo and bulk density (Busch et al. 2007b).

DA has a low-amplitude lightcurve, suggesting that the asteroid is not particularly elongated. The radar echoes have a nearly constant bandwidth and visible range extent, supporting this conclusion. The rotation period, 2.12160 ± 0.00004 h, is near the theoretical limit for a strengthless asteroid; in the NEA population, with two known exceptions (2001 OE84, 0.486 ± 0.002 h, Pravec et al. 2002, and 2001 VF2, 1.393 ± 0.001 h, Whitely et al. 2002), only objects larger than about 200 m rotate more rapidly (Pravec 2010, Harris & Warner 2010). Based on a smooth, red-trending spectrum and the lack of any detectable thermal emission, Rivkin et al. 2005 assigned 1950 DA a minimum optical albedo of roughly 0.2 and a taxonomic class of E or M.

DA has SC/OC = 0.14 ± 0.03, one of the lowest known for a near-Earth asteroid (Benner 2008): DA's surface is very smooth at centimeter to decimeter scales, comparable to Phobos and Deimos, suggesting that it is not an E-type. The delay-Doppler images show relatively modest surface topography, with the exception of one prominent angular feature observed at both Arecibo and Goldstone (Fig. 3.12, Appendix 1.b). In the high-resolution Arecibo images, there are obvious concavities and ridge-like features.

I find two possible pole directions and corresponding models, with estimated standard errors of 5° for each pole. The two models provide comparably good fits to the radar and lightcurve data and differ in pole direction by 165°; one is prograde and the other is retrograde (Table 3.3, Fig. 3.13). The two models have very different shapes and surface properties. The prograde model is very slightly elongated, with angular, perhaps faceted, relief and obvious indentations. The retrograde solution is

30

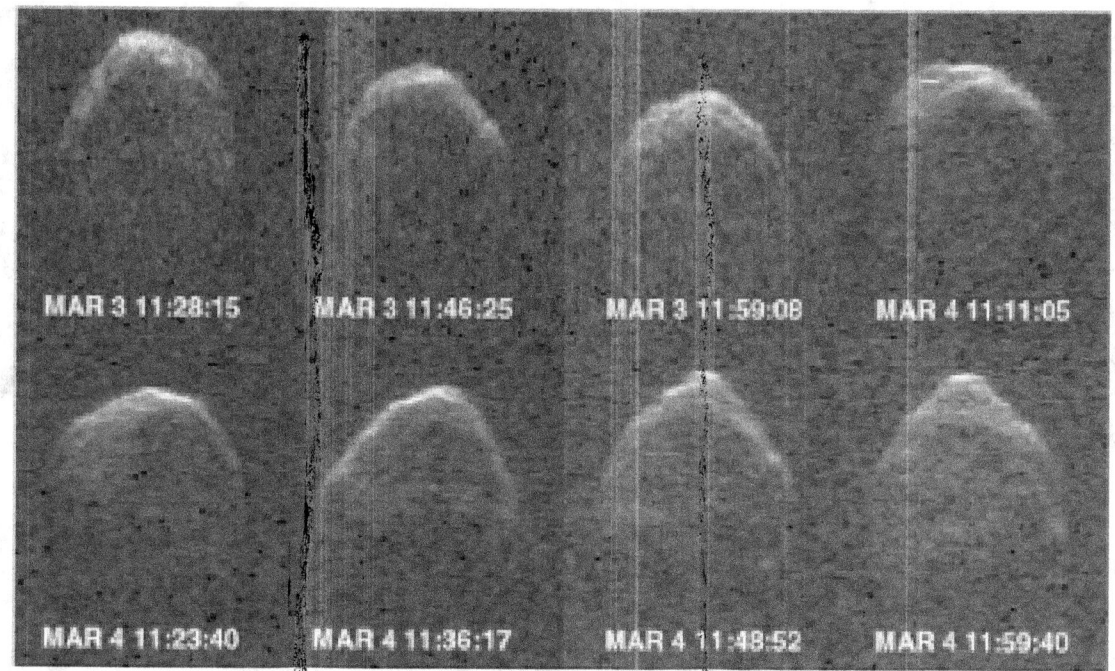

Figure 3.12: Selected delay-Doppler images of 1950 DA from Arecibo in 2001, with resolution 0.1 μs (15 m) x 0.256 Hz, showing the relatively constant bandwidth and angular surface features. For all radar images and corresponding model fits for both potential pole directions, see Appendix 1.b. Figure from Busch et al. 2007b.

Table 3.3: 1950 DA Model Properties

	Prograde Model	Retrograde Model
Pole Direction (λ, β), ±5º	(88.6°, 77.7°)	(187.4°, −89.5°)
DEEVE dimensions (2a, 2b, 2c) (±10%):	1.16x1.17x1.15 km	1.39x1.46x1.07 km
Maximum extents along principal axes (±10%):	1.19x1.28x1.24 km	1.45x1.60x1.20 km
Diameter of Equal-volume sphere (±10%):	1.16 km	1.30 km
Volume (±30%)	0.82 km³	1.14 km³
Mean OC Radar Cross-section:	0.36 ± 0.03 km²	
Polarization Ratio:	0.14 ± 0.03	
OC Radar Albedo:	0.35 ± 0.07	0.23 ± 0.05
V-Band Optical Albedo:	0.25 ± 0.05	0.20 ± 0.05
Minimum Strength-less Bulk Density (±10%):	3.0 g/cm³	3.5 g/cm³
Minimum Surface Grain Density (±10%):	3.2 g/cm³	2.4 g/cm³

Properties of the two 1950 DA models. V-band albedo was obtained using absolute magnitude H = 12.5. The minimum strength-less bulk density assumes inward acceleration at surface >= 0. The surface grain density is constrained by the radar albedo.

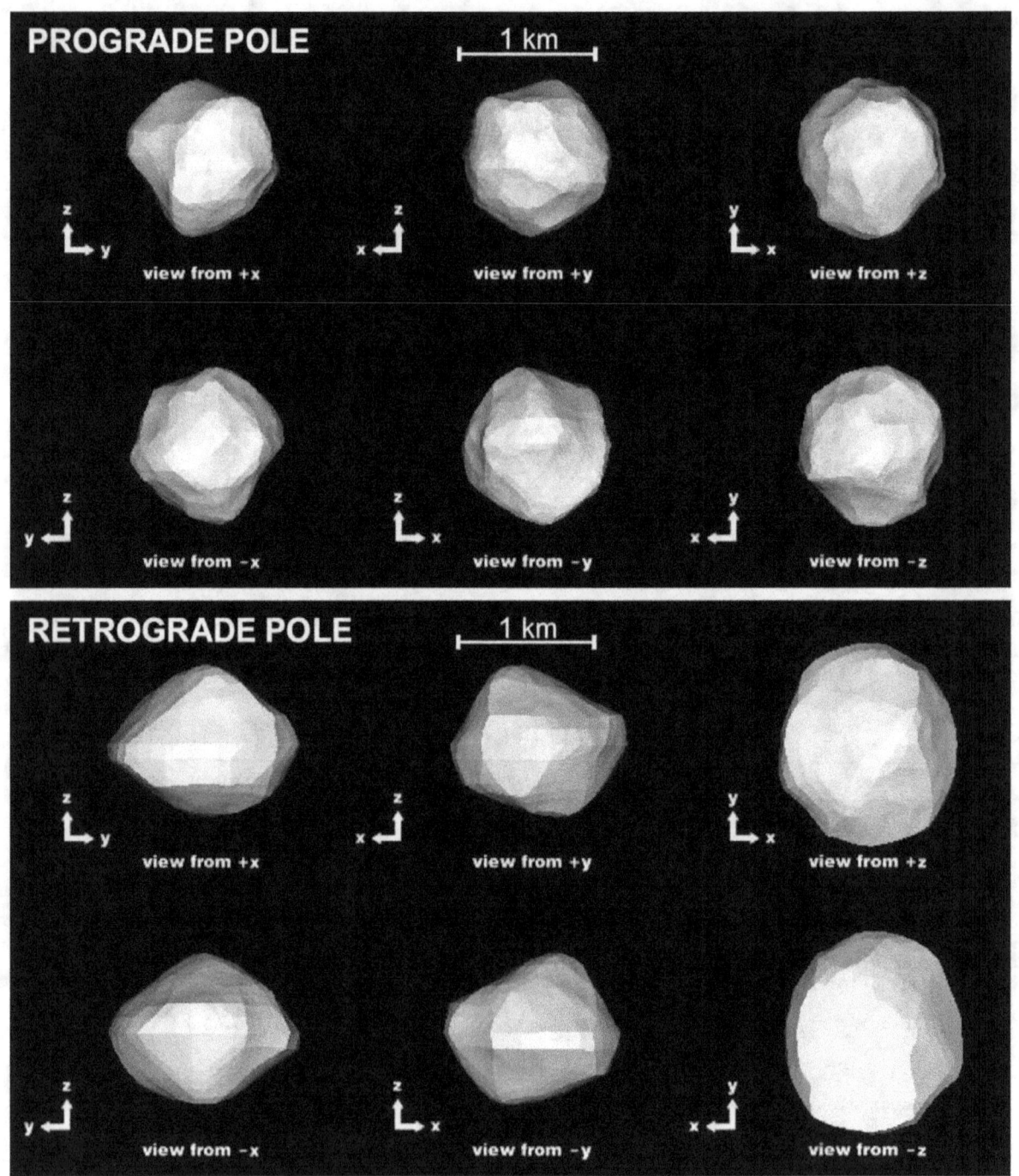

Figure 3.13. Principal-axis projections of the prograde (top) and retrograde (bottom) models of 1950 DA. Yellow shading marks regions that were not seen at incidence angles less than 75° (due to the echo strength, higher incidence angles were useful than for EV5 or SK).

quite oblate, with a prominent equatorial ridge. The retrograde solution implies higher sub-radar latitude and requires a larger size to match the observed bandwidth.

Because of their different cross-sectional areas, the prograde model has a radar albedo of 0.35 while the retrograde model has an albedo of 0.23. Both values are higher than those reported for most other asteroids, although not as high as for 1986 DA (0.58) or 216 Kleopatra (0.60), where the radar albedo estimates force the inference of a metallic composition (Ostro et al. 2002). The minimum surface grain densities (assuming zero porosity) are 3.2 g/cm^3 for the prograde model and 2.4 g/cm^3 for the retrograde model.

The surface grain density further constrains DA's composition. Based on the E- or M-type classification, there are three major compositional possibilities: enstatite achondrite, enstatite chondrite/hydrated silicates, and nickel-iron. Enstatite achondrite is excluded by the low polarization ratio. If 1950 DA has a density comparable to enstatite chondrites, ~3.6 g/cm^3, then it must have a low surface porosity (<30%); that is, it must be nearly lacking in regolith. If 1950 DA is metal-rich (analogous to iron meteorites), then the surface porosity must be less than ~60% for the prograde model and less than ~70% for the retrograde model; that is, the asteroid's regolith porosity is near or within the range seen for the Moon, 30-55% (Carrier et al. 1973).

For the retrograde model, if I assume a uniform bulk density of 2.5 g/cm^3, there is a large zone around the equator where rotational acceleration is greater than gravitational attraction. Increasing the bulk density of the retrograde model to 3.5 g/cm^3 just removes this zone. While even rubble pile asteroids have some tensile strength (Holsapple 2004), if DA had significant tensile strength, I would expect the topography to be more rugged, while the retrograde model is relatively smooth. Regardless of its rotation state, DA is likely denser than 3.5 g/cm^3, and either nickel-iron or a mixture of nickel-iron and rock.

3.e. Spin state and trajectory prediction

The surface temperature distribution and the magnitude and direction of the Yarkovsky acceleration are determined by an asteroid's shape and surface structure, but most importantly by the spin state. Delay-Doppler shape modeling frequently suffers from mirror-ambiguities, and this prevents accurate trajectory prediction. The shape models of 1950 DA and 2008 EV5 illustrate this problem.

3.e.i. 1950 DA in 2880

During the course of the radar observations, DA's orbit and ephemeris were progressively updated. This radar astrometry revealed an extremely close Earth approach on March 16 2880, with a probability of impact initially estimated as ~1/1000, which had not been foreseen despite the fifty-year optical astrometric arc. Further analysis of the predicted close approach revealed the importance of non-gravitational perturbations, particularly the Yarkovsky acceleration, on DA's trajectory (Giorgini et al. 2002). For an object of DA's size and approximate density, the Yarkovsky-produced offset in position in 2880 is of the order of 3000 Earth radii – large in comparison to all other sources of uncertainty, which accumulate to roughly 300 Earth radii (dominated by the gravitation perturbations during planetary and asteroid flybys). Given the uncertainty in the Yarkovsky perturbation when DA's shape and pole direction were unknown, Giorgini et al. 2002 concluded "the maximum probability of impact is best expressed as being between 0 and 0.33%." This ambiguity in pole direction and impact probability was the motivation for my shape modeling.

I currently have no basis for preferring one of my DA models to the other. If the rotation is retrograde, then the sum of non-gravitational forces, perturbations from other asteroids, and 15

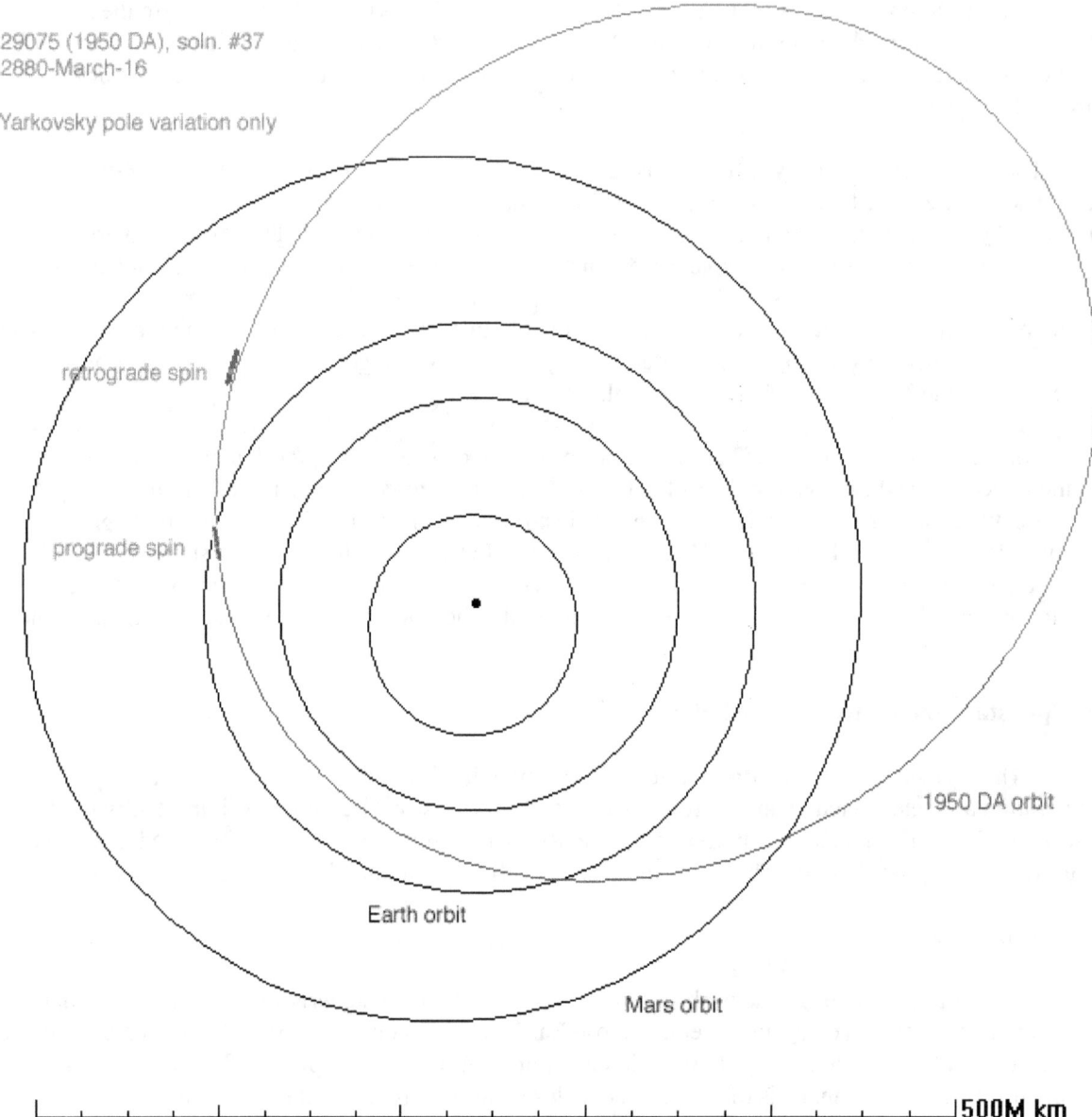

Figure 3.14: Predicted position of 1950 DA on 2880 March 16 for the prograde and retrograde models, showing the approximate effect of Yarkovsky on its trajectory – the magnitude of the offset depends on DA's actual mass. View is from ecliptic north. The orbit plotted is based on the instantaneous orbital elements of the prograde trajectory solution at that time. Earth is at the intersection of the prograde spin uncertainty region and the planet's orbit. Plot courtesy Jon D. Giorgini.

encounters with Earth and Mars at distances less than 0.1 AU, will alter the trajectory and delay the highest-probability arrival time at the intersection with Earth's orbit in 2880 by more than two months. This would eliminate the potential impact (Fig. 3.14). Alternatively, if the prograde model is correct, then the potential impact remains possible and a more extensive statistical analysis of potential perturbations due to more than 100000 known asteroids would be required for an improved impact probability estimate.

DA's next close Earth approach able to yield current-Arecibo SNRs high enough for ranging is in 2032. However, because the asteroid's sky position will be close to the sky arc of the 2001 radar experiment, there will not be any additional leverage on the spin state from radar imaging. The Yarkovsky perturbation will not be detectable in 2032, but will be detectable with current radar systems during the asteroid's 2074 Earth approach.

YORP analysis for DA implies a rate of change of the period of order $\sim 10^{-4}$ s/yr for both models. The change in period accumulates to roughly 0.1 s between now and 2880, and does not significantly affect calculations of impact probability. Estimates for YORP modification of the pole direction through 2880 are much smaller than the current uncertainties, so the models cannot be distinguished on the basis of thermal re-radiation effects for several decades, even if the YORP predictions were not very uncertain (Statler 2009). Optical lightcurves may be able to distinguish between the pole directions before the 2032 approach: the models predict very different lightcurve amplitudes in 2012 and in 2032.

Regardless of when the ambiguity in DA's spin state is resolved, it is a dramatic illustration of the need for alternative and unambiguous techniques to determine asteroid spin states. As the number of known asteroids increases, so will the number of objects with potential impacts. The ability to estimate the effects of Yarkovsky and YORP will be critical to reliable predictions.

3.e.ii. 2008 EV5

EV5 was discovered on March 5 2008, by the Mt. Lemmon survey. It was observed briefly before moving too far from the Earth to be observed. Robert McNaught of Siding Spring observatory in Australia recovered it in late November 2008, one month prior to the Earth approach. EV5 was a Goldstone delay-Doppler radar target on 2008 December 16-21, an Arecibo delay-Doppler target on 2008 December 23-27, and a VLBA+GBT multiple-station target on 2008 December 23 (Busch et al. 2010a, 2010b).

The EV5 delay-Doppler images have 0.125-µs resolution in time delay (19 m in range) at Goldstone and 0.05-µs (7.5 m) at Arecibo. While the Goldstone images have lower resolution, they constrain EV5's shape and provide leverage on its pole direction: over the course of the Goldstone tracks, the echo bandwidth changed by ~15%, due entirely to changes in sub-radar latitude. The delay-Doppler images from all four days at Arecibo are quite strong and show prominent structures on EV5's surface (Fig. 3.15, Appendix 1.c, Fig. 3.5). Subject to the limitations of SHAPE in determining the size of the equatorial ridge (Sec. 3.d, Fig. 3.4), I conclude that EV5 is roughly spheroidal, consistent with the low lightcurve amplitude, but it has significant surface topography. There are both depressed areas and an equatorial ridge (seen best on 2008 Dec 26). EV5's shape is similar to those of several other asteroid radar targets (e.g. 1998 KW4, Ostro et al. 2006; 1998 CS1, Benner et al. 2009; 2006 VV2, Benner et al. 2007b) and suggests a rubble-pile structure.

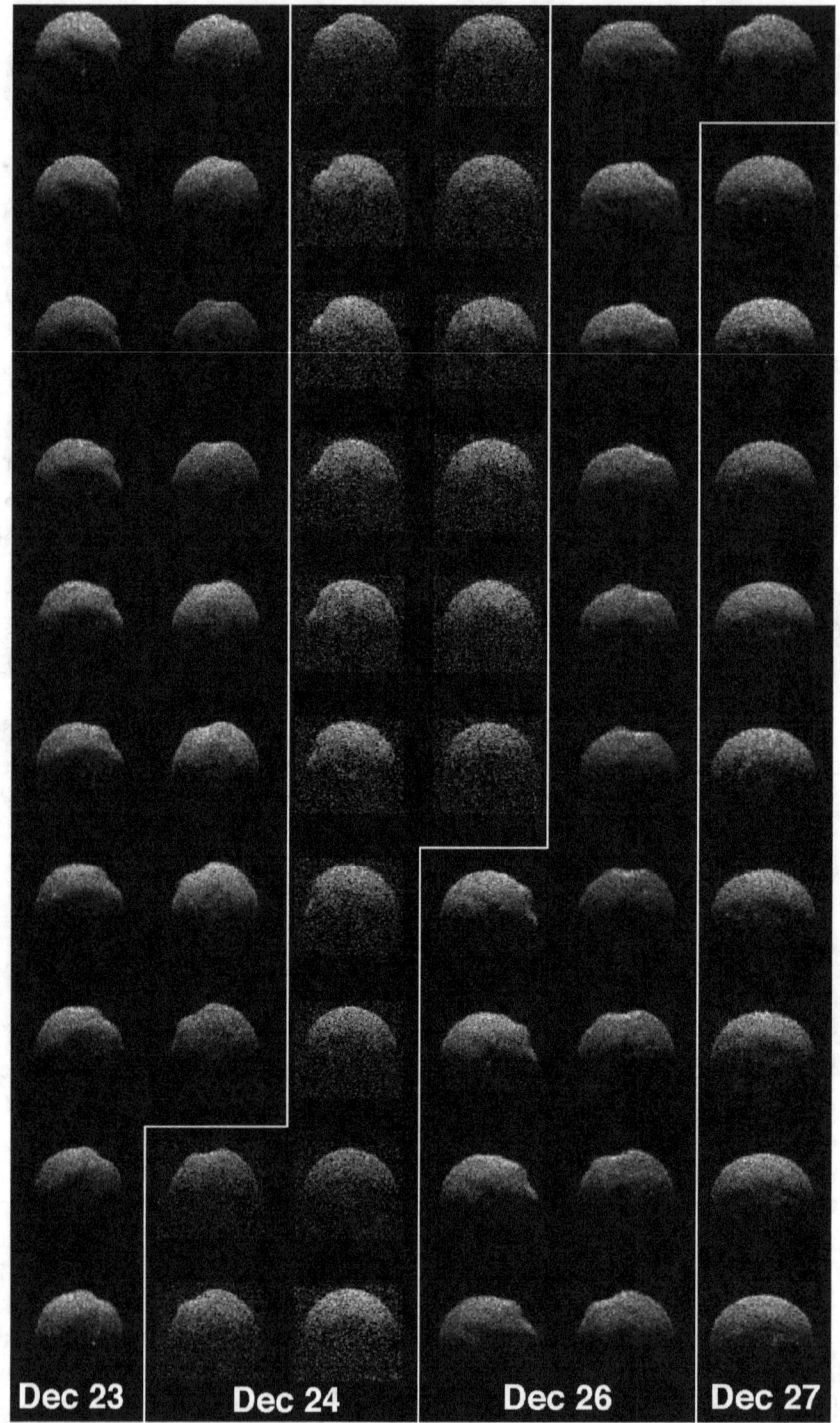

Figure 3.15: Selected delay-Doppler images of 2008 EV5, showing the relatively circular leading edge broken by the concavity. These are Arecibo radar images taken in December 2008. Resolution is 0.0625 Hz x 0.05 μs (7.9 mm/s x 7.5 m). In the collage, time increases from top to bottom and left to right. For a full collage, including both Arecibo and Goldstone data and corresponding fits, see Appendix 1.c. Figure from Busch et al. 2010b.

Table 3.4: 2008 EV5 Model Properties

Rotation Period: 3.725±0.001 h (Galad et al. 2009)
Spectral Type: C (M.D. Hicks, pers. comm.)
Absolute Magnitude H: 20.0

Pole Direction, ecliptic: (276º, −74º) ± 10º
Maximum dimensions along principal axes: (490 x 490 x 480) ± 40 m
Equivalent diameter: 450 ± 40 m
Volume: 0.052 km^3 ± 30%

Polarization Ratio (SC/OC): 0.38 ± 0.03

OC Radar Albedo: 0.25 ± 0.06
Optical Albedo: 0.09 ± 0.03

Properties of the 2008 EV5 shape model, for the retrograde pole direction. The model is a polyhedron with 2000 vertices. Other than the pole direction (offset by 180º) and the detailed shape, the prograde shape has the same properties (maximum dimensions, equivalent diameter, volume, and albedos) to within the model uncertainties. The uncertainty in the model's z-axis extent and the overall height of the ridge may be underestimated (recall Fig. 3.4 and the effects of changing penalty weights).

The delay-Doppler images of EV5 cover all longitudes, but only low temperate to equatorial latitudes. Consequently, they permit two possible pole directions and corresponding shapes – mirror solutions (Table 3.4, Fig. 3.16). Both nominal EV5 shapes are close to 450 ± 40 m spheres, with a low-latitude ridge broken by a prominent concavity.

Since the observations were all near the equator, exact location of the ridge is not well constrained: relatively large changes in latitude result only in small changes in distance from the Earth. The ridge may very well lie precisely on the equator, as observed on 1999 KW4 (Ostro et al. 2006). In any event, the large depression does break the line of the ridge, implying that it post-dates the ridge's formation. The concavity is ~150 m across and ~30 m deep, perhaps suggestive of a large impact crater and reminiscent of those seen on 52760 (1998 ML14) (Ostro et al. 2001) and WT24.

The ridgeline in the EV5 models is less than one facet wide and oriented exactly east-west, except where it is broken by the depression, producing a circle of alternately tilted facets around the object. This is a known artifact of SHAPE (Sec. 3.2.i, Magri et al. 2007). The true ridgeline is presumably somewhat irregular. EV5's radar echo indicates a normally rough surface. SC/OC is 0.38 ± 0.03, above average but not dramatically so (Benner et al. 2008). EV5 has radar albedo 0.25 ± 0.06 and optical albedo 0.09 ± 0.03. The radar albedo is consistent with a maximum near-surface bulk density of 3.2 g/cm^3 (Magri et al. 2001). Such a grain density, combined with the optical albedo, is consistent with a wide range of silicate-carbonaceous mixtures and low porosity.

EV5's shape is intriguing, particularly that the concavity bisects the ridge. If the concavity is indeed an impact crater, seismic shaking during the collision that formed it might be expected to destroy the ridge and the other smaller surface features (Asphaug 2008). It is possible that EV5's internal structure is very efficient at damping seismic waves (e.g. a relatively fine-grained rubble pile with a few

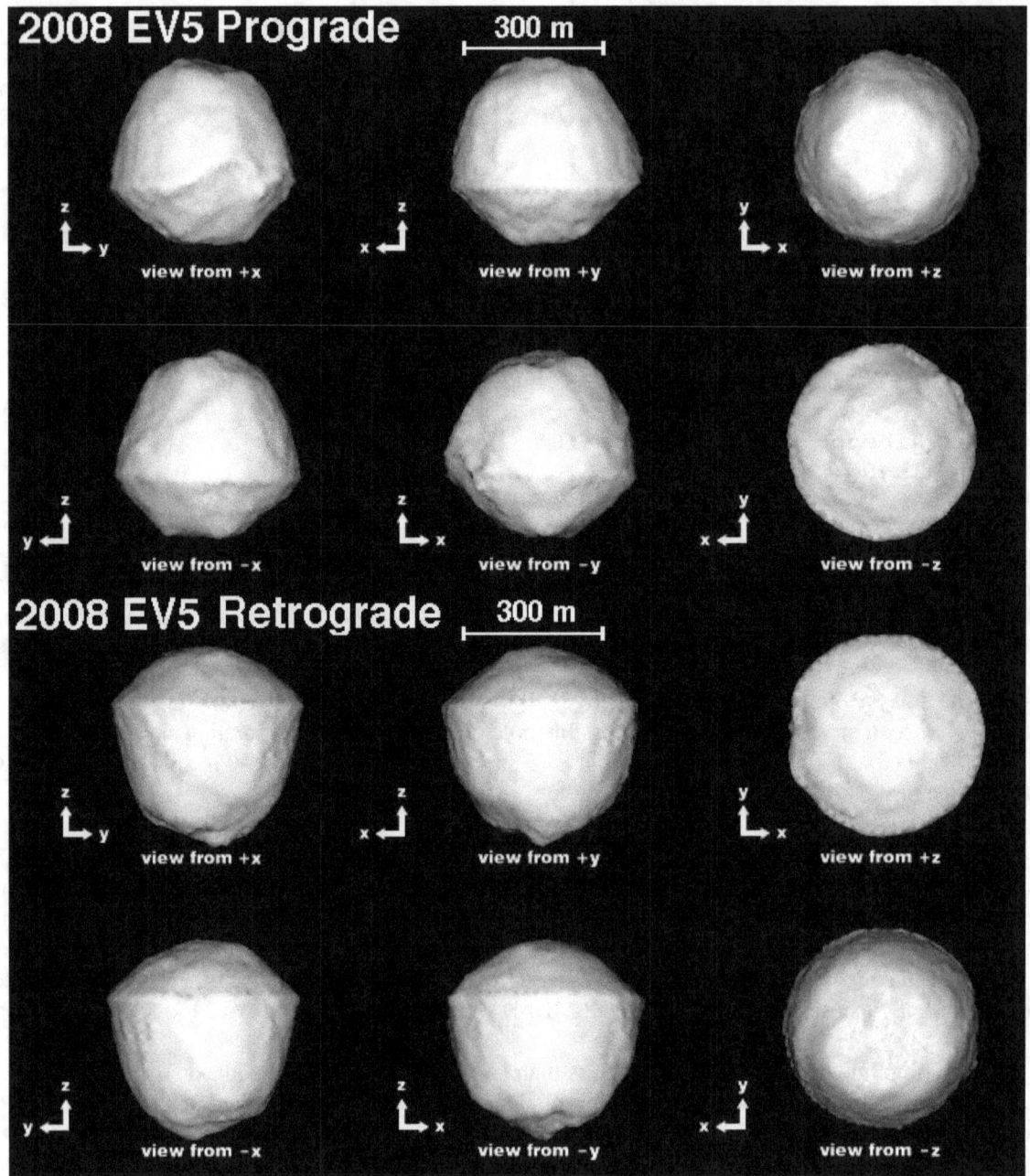

Figure 3.16. Nominal prograde (top) and retrograde (bottom) models of 2008 EV5. Figure from Busch et al. 2010b.

large blocks mixed in). This would ensure that the effects of an impact were minimal except immediate to the crater, but is a purely speculative conclusion.

The equatorial ridge is consistent with an episode of reconfiguration when EV5 was spinning faster. This suggests that YORP should currently be slowing down the rotation. For a range of thermal properties, both nominal models, as well as the large ridge retrograde model (Fig. 3.4), are indeed slowing down and were spinning at breakup velocity less than ~1 Myr ago, and would have been reconfigured at that time. However, the little ridge model predicts that EV5's rotation should currently be *accelerating*. I cannot specify EV5's shape well enough to determine if its rotation is speeding up or slowing down. Because it is so nearly spherical, even small changes in EV5's shape (e.g. facet-scale lumps in the models, Statler 2009) change the predicted magnitude and sign of the YORP torque. Regardless of EV5's pole direction or true shape, spin changes will have to be measured directly. The formation of the concavity will also have changed the torque, so a simple spin-up and spin-down is not a good approximation to EV5's rotation history over the last Myr, as with WT24.

The two potential pole directions give approximately opposite Yarkovsky acceleration for EV5. While there are no known potential impacts for either pole, the Yarkovsky offset between them accumulates to ~6 million kilometers over the next 170 years (Busch et al. 2010b).

In one sense, the ambiguity in the spin state of 2008 EV5 is fortunate. I had selected it to test new multi-station techniques of determining pole directions (Chap. 4). I was successful in determining the sense of EV5's rotation by the technique of radar speckle tracking: it rotates retrograde. By splitting the pole direction ambiguity, the uncertainty in EV5's trajectory is reduced by over a factor of 2, to ~2.5 million km 3-σ in 2180, limited by the residual uncertainty in the astrometry. EV5 demonstrates the progression that is now possible for future radar targets, using delay-Doppler imaging to obtain shape information in combination with new techniques to determine pole directions.

4. Ways to Determine Asteroid Pole Directions

Having seen the capabilities and limitations of asteroid radar – the combination of <10 m spatial resolution with frequently ambiguous pole direction estimates – I sought other ways to image asteroids. As described below, no other ground-based observing technique equals the resolution of delay-Doppler radar imaging for near-Earth asteroids. My goal was to obtain unambiguous measurements of pole directions and then to combine these with shape and morphological information from delay-Doppler data.

I have considered four separate techniques, two using the natural emission from the target objects and two using radar. In this chapter, I discuss optical and near-IR adaptive optics; imaging the radar echo of an object using the techniques of radio interferometry; and finally the technique of radar speckle tracking, which I have developed and applied to split pole direction ambiguities. My application of speckle tracking is described in Chap. 5. The final observing technique is to image asteroids' sub-millimeter thermal emission. This requires significant new hardware, and is discussed in Chap. 6.

4.a. Adaptive optics

Optical telescopes are normally limited to angular resolution of ~1° by atmospheric seeing. Such resolution is insufficient to image asteroids. This is reflected in the word 'asteroid', which means 'star-shaped' and was originally coined by William Herschel to refer to any object in the solar system that was an unresolved point (Herschel 1802). In consequence, there is a long history of estimating asteroids' rotation periods, pole directions, shapes, and thermal properties from the variation in disc-integrated brightness with time - asteroid lightcurves (e.g., Kaasalainen et al. 2002).

As described in Chap. 3, lightcurve-derived rotation period measurements are very precise and essential to radar observations. Lightcurve-derived pole directions provided the first demonstration of the importance of YORP, but usually have uncertainties of 20° or more and often are ambiguous, particularly for objects that are not elongated (as demonstrated by 1950 DA and 2008 EV5). Finally, lightcurve estimates of asteroid shapes are not sensitive to concavities and provide at best an estimate of an object's convex hull, analogous to wrapping the asteroid in plastic sheeting (Kaasalainen et al. 2002). A convex hull shape model does not yield an accurate estimate of Yarkovsky or of YORP.

While it is easy to obtain lightcurves for a very large number of objects at optical and infrared wavelengths, without spatial resolution they are of only limited use. Near-Earth asteroids have been marginally resolved using ground-based speckle interferometry (Drummond et al., 1985) and deconvolution of images obtained from the 2.5-m Hubble space telescope (Noll et al., 1995). The former is limited by sensitivity and the latter by resolution. More recently, adaptive optics (AO) systems on 10-m ground-based telescopes have improved their resolution by more than an order of magnitude. The AO on the Keck II telescope now routinely achieves <40 mas resolution (Wizinowich et al. 2000), sufficient to resolve and image the largest main-belt asteroids (Marchis et al. 2006b) and occasional near-Earth objects (Fig. 4.1, Busch et al. 2007c).

AO is a particularly powerful tool to locate asteroids' satellites, or to set upper bounds on their existence (Busch et al. 2009). The majority of multiple main belt asteroids have satellites only a few kilometers across (Marchis et al. 2006b). However, almost all binary near-Earth asteroids are too small and/or too close to each other to be imaged with AO, and have been discovered by radar (Margot et al.

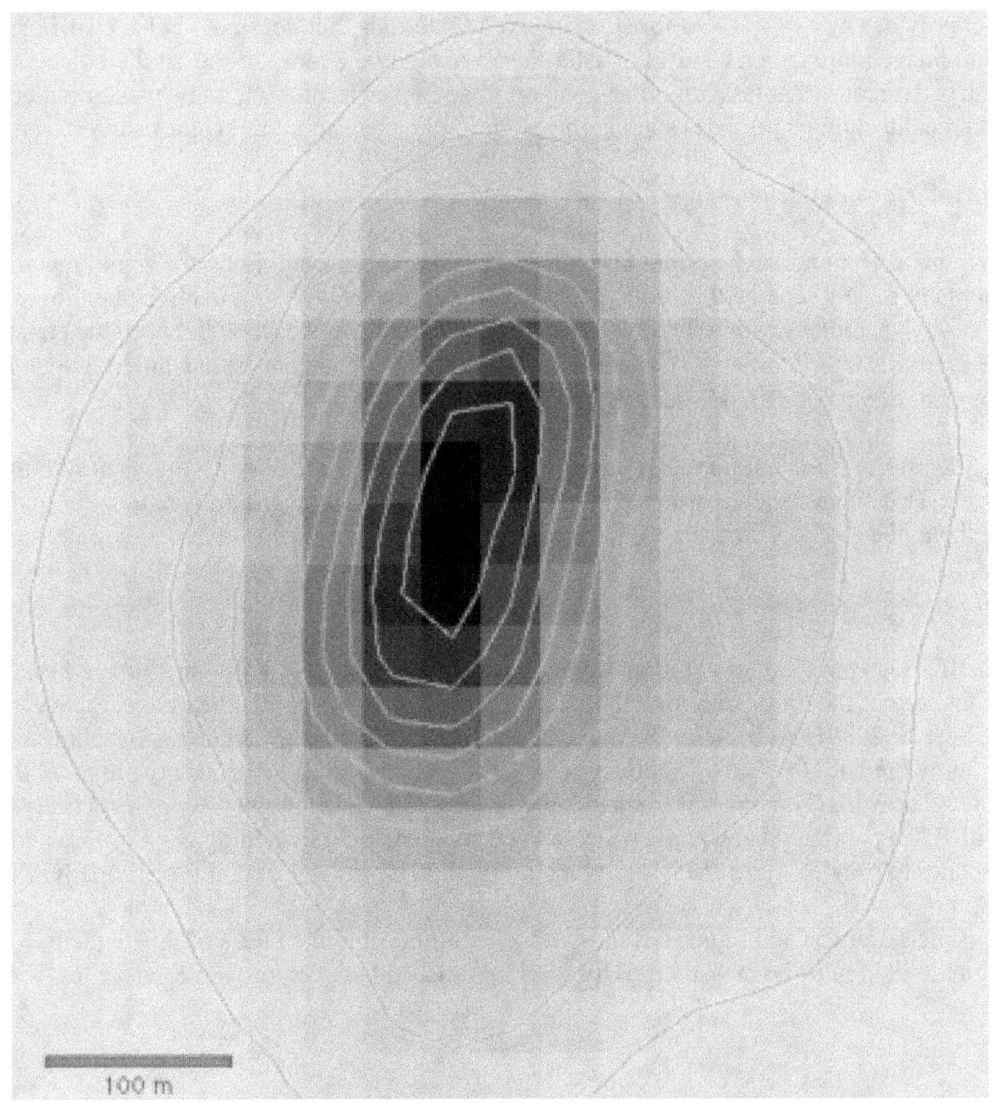

100 m

Figure 4.1: Keck H-band (1.65 μm) adaptive optics image of the near-Earth asteroid 2004 XP14, during its 1.1 lunar distance close Earth approach on 2006 July 3, 12:55 UTC. The integration time was 5 s. Image scale is 33 m/pixel. The spatial resolution of the image is 0.040" = 132 m at the object. Contours are spaced linearly between 10% and 90% of the peak flux. North is up and east to the left. Solar illumination is from 20° south of east. Figure from Busch et al. 2007c.

2002). As Fig. 4.1 shows, only the closest-approaching near-Earth asteroids are more than 40 mas across. 2004 XP14 passed at 1.1 lunar distances (440000 km) from the Earth in 2006 (Benner et al. 2007a, Busch et al. 2007c) and had best-fit plane-of-sky ellipse dimensions of (300±65)×(300±120) m. This gave it the largest angular size of any known near-Earth object during the four-year period from 2005 to 2008. In 2008, the objects Toutatis and 2008 TC3 both exceeded 40 mas in angular size – the latter only immediately before impacting the Earth's atmosphere and forming a fireball over northern Sudan (Jenniskens et al. 2009).

AO provides a powerful tool for studying some asteroids, but does not yet have sufficient resolution to image more than a handful of asteroids or to study the morphology of the objects that can be resolved. That requires at least an order-of-magnitude higher resolution, a few mas or better. Only one astronomical technique provides such high-resolution images: aperture synthesis interferometry.

4.b. Radar interferometric imaging

In radio astronomy, high resolution images are obtained by combining the signal recorded from a source at many widely separated stations, to synthesize an image with an effective plane-of-sky resolution of λ/D_{max} radians, where D_{max} is the maximum distance or baseline between the stations. This is aperture synthesis interferometry. Longer baselines provide higher spatial resolution, which has led to the development of very-long baseline interferometry (VLBI).

While asteroids are not normally strong radio sources, they are very bright when illuminated by a radar beam. This suggests a possible technique to image radar targets: illuminate them with the radar and image them plane-of-sky with a VLBI array, effectively at right angles to delay-Doppler imaging.

4.b.i. VLBI Arrays

VLBI arrays have antenna stations spread across hundreds to ~10000 km, and are normally limited by the size of the Earth, although there have been experiments with antennas on spacecraft (Hirabayashi et al. 2000). At present, the most active VLBI array is the Very Long Baseline Array (VLBA), run by the NRAO (Fig. 6.1, Napier et al. 1994), which I used for all of my radar-VLBI observations. The VLBA operates at frequencies between 300 MHz and 90 GHz and contains 10 stations with 25-m paraboloid antennas, which can be supplemented by adding one or more additional radio telescopes to provide higher sensitivity, including the 100-m Green Bank Telescope (GBT).

For interferometry to be successful, both the amplitude and the phase of the incident electromagnetic field must be measured. This is accomplished by a heterodyne receiver that splits the

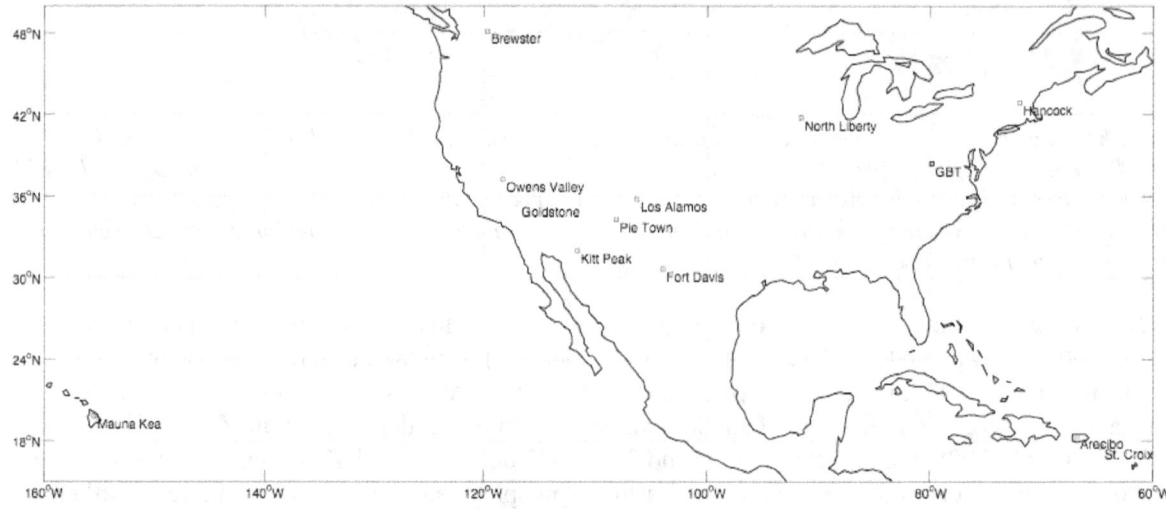

Figure 4.2: A representative interferometric array: the stations of the Very Long Baseline Array (VLBA), including the Green Bank Telescope. Also plotted are the Arecibo and Goldstone radar sites.

electric field into orthogonal polarizations, down-converts the signals from the antenna feed to a lower frequency, filters out a useable bandwidth, and then samples the antenna receiver voltages and digitizes them for analysis. The VLBA currently can sample bandwidths between 62.5 kHz and 16 MHz in up to 16 simultaneous channels, with quantization of 1 or 2 bits (Napier et al. 1994). Because the VLBA stations have not been connected directly, the voltage samples are timestamped and recorded to discs, which are shipped to the array center in Socorro, New Mexico, for processing. The raw voltage data are synched in time and used to reconstruct the brightness distribution of the sky (an image).

4.b.ii. Definition of Visibility

To convert the raw voltage data from each antenna into information about the brightness distribution of the sky, aperture synthesis considers the antennas in pairs, of which there are N*(N-1)/2 in an array of N elements, e.g. 55 for the 11-station VLBA+GBT.

The receiver voltage measured at a station is proportional to the magnitude of the electric field of a particular polarization received at the antenna. Following Taylor et al. 1999, if the sky is a radiator at some very large distance R, and the station is at some position $\vec{\mathbf{r}}_i$, then the voltage received by a station i is:

$$(4.1) \qquad V_i \propto \int_S G_i(\sigma)E(\sigma)\frac{e^{-\frac{2\pi i v |\vec{\mathbf{R}}(\sigma)-\vec{\mathbf{r}}_i|}{c}}}{\left|\vec{\mathbf{R}}(\sigma)-\vec{\mathbf{r}}_i\right|}d\sigma$$

where σ is an element of surface area on the sky (S); $\vec{\mathbf{R}}(\sigma)$ is the vector in the direction of σ from the origin with length R; G_i is the beam pattern of the antenna; E is the electric field in one polarization radiating along the vector from a particular point on the sky, and v is the observation frequency. Eq. (4.1) is the equation for the electric field of a traveling light wave integrated over the sky, weighted by the beam pattern. Note that $\vec{\mathbf{r}}_i$ changes with time, as the Earth rotates relative to the sky.

A VLBI array combines the signals from a pair of antennas by *correlating* the voltage streams. Consider the cross-correlation between the electric field at stations i and j: the expectation of the complex product $\langle V_i V_j^* \rangle$. The cross-correlation is equal to the integral of the product of the integrand of Eq. (4.1) for the two stations:

$$(4.2) \quad \langle V_i V_j^* \rangle \propto \int_S \int_S G_i(\sigma_i)G_j(\sigma_j)E(\sigma_i)E^*(\sigma_j)\frac{e^{-\frac{2\pi i v |\vec{\mathbf{R}}(\sigma_i)-\vec{\mathbf{r}}_i|}{c}}}{\left|\vec{\mathbf{R}}(\sigma_i)-\vec{\mathbf{r}}_i\right|}\frac{e^{\frac{2\pi i v |\vec{\mathbf{R}}(\sigma_j)-\vec{\mathbf{r}}_j|}{c}}}{\left|\vec{\mathbf{R}}(\sigma_j)-\vec{\mathbf{r}}_j\right|}d\sigma_i d\sigma_j$$

For a source that is spatially incoherent, the product of the two electric fields will average to zero except when the two σ's refer to the same location on the sky. The integral can be condensed and the exponential terms combined. Making the further assumption that the area of sky occupied by the source being imaged is small and therefore can be approximated as planar, Eq. 4.2 becomes:

$$(4.3) \quad \langle V_i V_j^* \rangle \propto \int_S G_{ij}(\sigma)\langle E(\sigma)E^*(\sigma)\rangle\left|\vec{\mathbf{R}}(\sigma)\right|^2 \frac{e^{\frac{2\pi i v \vec{\mathbf{b}}_{ij}\cdot\vec{\sigma}}{c}}}{\left|\vec{\mathbf{R}}(\sigma)-\vec{\mathbf{r}}_i\right|\left|\vec{\mathbf{R}}(\sigma)-\vec{\mathbf{r}}_j\right|}d\sigma$$

Here the baseline vector $\vec{b}_{ij} = \vec{r}_i - \vec{r}_j$ is the component of the separation between the stations that is normal to the line-of-sight. $G_{ij}(\sigma) = G_i(\sigma)G_j(\sigma)$ is the product of the beam patterns of the town antennas.

Converting from the electric field to the sky brightness distribution $I(\sigma) \propto \langle E(\sigma)E(\sigma)^* \rangle$ and allowing R to go to infinity, means that all terms involving $\mathbf{R}(\sigma)$ cancel. This produces the fundamental equation of synthesis imaging: the definition of the *visibility* as the cross-correlation between the voltages measured at two of the antennas in the array. The visibility for stations i and j is defined as:

$$(4.4) \qquad V_{ij} = \langle V_i V_j^* \rangle = \int_S G_{ij}(\sigma)I(\sigma)e^{-\frac{2\pi i v \vec{b}_{ij} \cdot \vec{\sigma}}{c}} \, d\sigma$$

V_{ij} is of course a complex number, with an amplitude and a phase. For a sufficiently large magnitude D of $\vec{b}_{ij}$, a very small change in the plane-of-sky position produces a large change in the phase of V_{ij}. Physically, this is due to the difference in the path lengths from the source to the two receivers (Fig. 4.3.a).

For imaging, it is more important that each visibility – and each antenna pair at a given time – samples a single spatial frequency component of the Fourier transform of $I(\sigma)$, the plane-of-sky image, with the modification of $G_{ij}(\sigma)$. For VLBI applications, including asteroid radar, the sources that make up $I(\sigma)$ are very compact in angular extent and a very large distance from other compact sources. In this case, the beam pattern can be approximated as independent of σ. $G_{ij}(\sigma)$ becomes the product of two complex gain parameters g_i and g_j, where the gain has an amplitude and phase for each station at each epoch, and can be factored out of the integral. The spatial frequency sampled by a given visibility is determined by $\vec{b}_{ij}$ (Fig. 4.3.b). Technically, each pair of antennas samples a pair of frequencies, one the negative of the other, corresponding to the two possible orders of correlation V_{ij} and V_{ji}. However, these two visibilities are necessarily complex conjugates of each other (since $\langle V_i V_j^* \rangle = \langle V_j V_i^* \rangle^*$) and they are not distinct in information content.

For a larger D, the baseline vector $\vec{b}_{ij}$ samples a higher spatial frequency and correspondingly smaller scale structure in $I(\sigma)$. As an example, Fig. 4.3.c shows the amplitude of the visibility for a disc source as a function of baseline length (the visibility amplitude for a disc follows the Airy profile familiar from normal optics). As long as the gains are constant, the visibility amplitude is independent of the position of the source, but the visibility phase changes. Due to the dot product in the exponential in Eq. (4.4), only motion parallel to the baseline produces a phase difference (Fig. 4.3.d). The relationship between visibility phase and sky position also wraps in phase after a position shift of λ/D. One antenna pair cannot locate a source on the sky. But combining many pairs of antennas provides a large number of visibilities and allows locating the source and determining its structure - in other words, measuring $I(\sigma)$ - to approximately the diffraction limit of the longest baseline.

Equation (4.4) is given for a monochromatic visibility, at frequency v. In practice, this corresponds to one frequency channel of finite width. If that width is too large, then Eq. (4.4) is not exact, because different spatial frequencies have been blurred together. This is dealt with by assembling an image from multiple channels, which also allows exploration of any spectral dependence in $I(\sigma)$. For radar targets, the total echo bandwidths are low enough (fractional bandwidths are $\Delta v / v \leq 10^{-7.5}$) that a single frequency channel match-filtered to the echo will obey Eq. (4.4). However, since there is information in the frequency distribution of the echo (Sec. 3.1), spectral line imaging remains useful.

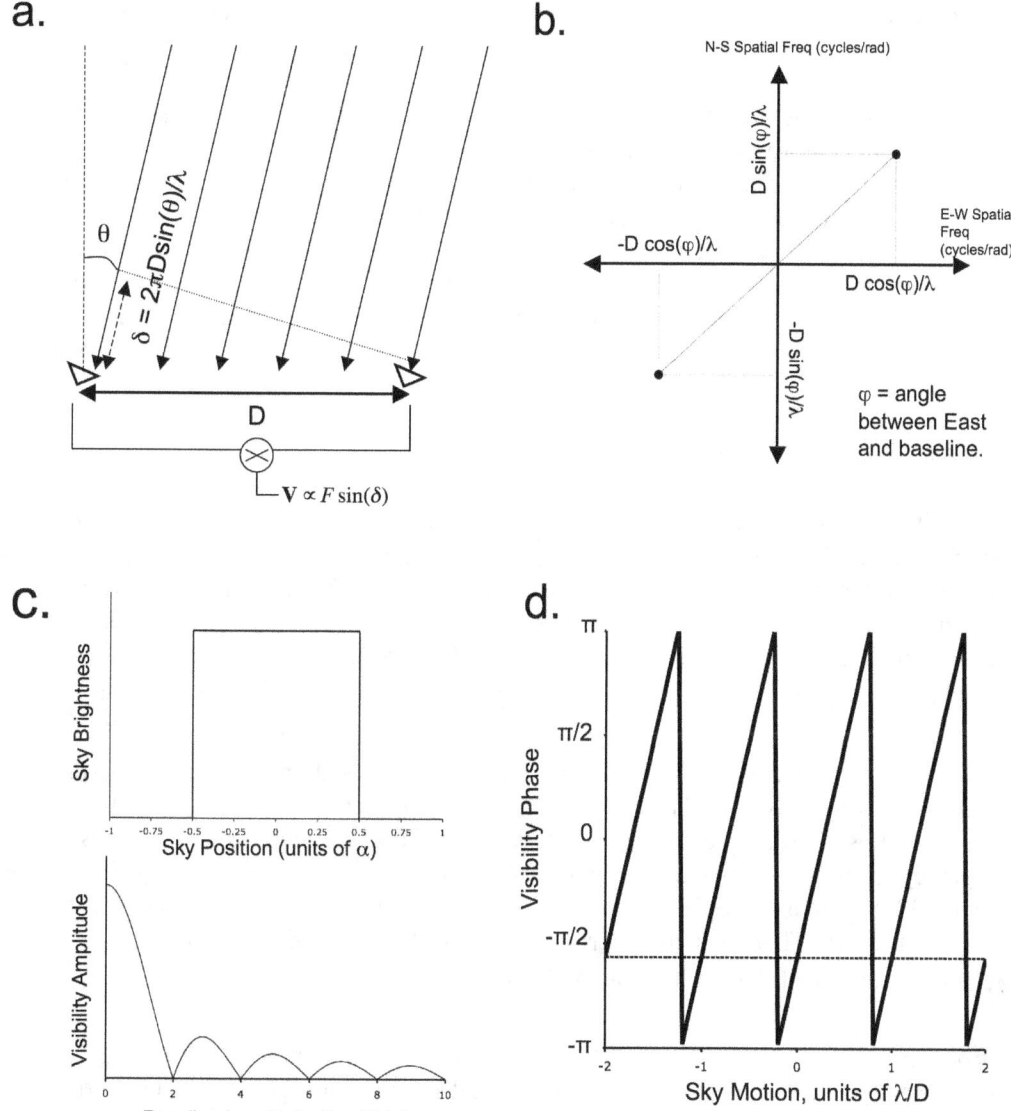

Figure 4.3: a. Generic two-element interferometer, with its response function for flux at a given angle. An array forms a collection of two-element interferometers, one for each pair of stations. b. Spatial frequency sampled by a pair of stations for a given relative separation (baseline). c. top – A uniform source of angular size α. bottom - Visibility amplitude for a pair of stations as a function of baseline length. d. Effect of moving the source from (c) across the sky – parallel (solid) and normal (dashed) to the baseline. Visibility phase changes when the source is moved parallel to the baseline, from an initial value determined by the source structure. For this example, the motion can be assumed small enough that antenna gain is constant, so the visibility amplitude remains the same.

A final assumption in Eq. (4.4) is that the source is in the far field of the instrument, i.e. the incident wavefronts are planar. This is not the case for a planetary target: the solar system is in the near field. The near-field distance is approximately:

$$(4.5) \qquad D_{nearfield} = 2D_{max}{}^2 / \lambda$$

Thus, for the VLBA (D_{max} = 8600 km), the near field distance at 2380 MHz (12.6 cm) is about 8000 AU. This has a major effect: in the near field, the curvature of the wavefront distorts the correlation between the stations, introducing phase errors as the signal arrives at slightly different times.

For a source at a known distance, appropriate corrections in the time delays can restore the correlation for spectral channels narrower than the reciprocal of the uncertainty in the corrections. Asteroid ephemerides are typically accurate to a few microseconds with delay-Doppler radar astrometry and the maximum bandwidth required is a few kilohertz, so near-field correction is possible. The near-field corrections are distinct from the geometric corrections necessary for a source at infinite distance, which are only equivalent to projecting the stations onto a plane perpendicular to the direction to the source. Near-field corrections can be considered equivalent to refocusing the array: they defocus any background sources and distort wide-field images at the same time as they correct the image close to the center (Taylor et al. 1999). This is not significant for asteroid targets milliarcseconds wide.

4.b.iii. Image Reconstruction from Visibility Data

Given accurately calibrated visibilities, the plane-of-sky image will be a best estimate of $I(\sigma)$. The cognitively simplest method is to take the spatial frequency component for each baseline, multiply it by the visibility, and add up the values for each pixel in an image. This makes a 'dirty map', which is the basis of any later deconvolution or imaging. This is the direct Fourier transform (DFT) method, although we are technically taking the inverse transform. The DFT is accurate, but also very slow: it scales as $N^2 n_{image}^2$, where N is the number of stations and n_{image} is the width of the map in pixels.

To greatly speed up runtime, consider all of the spatial frequencies at once, the uv-plane in interferometry terminology. Each visibility is equal to the value of the Fourier transform of the image at a point in EW (u) and NS (v) spatial frequency coordinates. By gridding the measured visibilities to uniform spatial frequency spacing, with zeros everywhere else, the Fast Fourier Transform (FFT) can be applied (e.g. Brigham 1974). The inverse FFT of the uv-plane populated only by the measured visibility points is of course the dirty map. There is some overhead in the gridding, but the FFT scales as $n_{image}^2 \ln^2(n_{image})$. It runs very much faster than the DFT for any significant N.

Gridding from specified uv points to a uniform grid requires consideration of weighting: a data point that does not lie exactly at a grid point must be interpolated to surrounding cells and tapered in a way that avoids sharp computational discontinuities. Visibilities are also typically weighted by length to adjust the shape of the synthesized beam to emphasize either point sources or extended structure. These considerations are discussed thoroughly elsewhere (Taylor et al. 1999).

4.b.iv. Limitations of Synthesis Imaging: Deconvolution and Source Modeling

The dirty map only contains the observed visibilities. To derive a best estimate of $I(\sigma)$, the missing uv-plane data must be filled in by some form of deconvolution, modeling the brightness distribution. For radio astronomy applications, the most commonly used method is CLEAN (Clark 1980), which models the image as a superposition of delta functions and successively removes them, providing a best-fit to the image if the source is not extended over an area many times the beam size and the signal-to-noise per beam is significantly greater than unity (Schwarz 1979, Cornwell et al. 1999). Unless the calibration of the data is exceptionally good for the VLBA, additional corrections to the antenna gains – both amplitude and phase – must be included, in a process called self-calibration (Pearson & Readhead 1984). While self-calibration decreases the amount of information in the image; in particular, absolute flux and absolute position are now unknown; it removes calibration errors that would otherwise be impossible to correct.

As an example of deconvolution and its limits, Fig. 4.4 shows a simple model (a frequency-independent uniform disc), the model uv-plane and the points in it sampled by a VLBA snapshot observation at two different frequencies, and corresponding dirty maps and CLEAN restorations. There are very obviously limits to deconvolution. If the object is too large (power at low spatial frequencies) and the number of visibilities sampled is too small, it is impossible to localize the object in space and the deconvolution will not converge to a physical result. For objects with sufficient uv sampling, deconvolution does produce accurate reconstructed images.

In radio astronomy, the Earth's rotation is used to sample additional visibilities: as the baselines change in orientation, they sample different uv points. However, almost all asteroid radar targets rotate quickly enough that this technique does not work. For a typical radar target with $P \approx 4$ h, $R \approx 0.5$ km, $d \approx 0.04$ AU, and maximum projected baseline ~8000 km, $t_{int} = 8$ min at 2380 MHz and 2.2 min at 8560 MHz. The Earth's rotation corresponds to a change of ~0.04 radians in the positions of the uv points in 8 minutes, which does not significantly improve the uv-plane coverage as compared to a snapshot.

4.b.v. Conditions for radar-interferometric imaging

The upper limit on the complexity of a reconstructed image is the number of degrees of freedom of the inversion model. Each visibility measurement provides an amplitude and a phase. The number of degrees of freedom is twice the number of visibilities minus calibration parameters. Since there are $(N*(N-1))/2$ visibility measurements for a snapshot, for self-calibrated data when the asteroid has been match-filtered and occupies one spectral channel, there are $(N*(N-1) - 2N)*$(number of epochs) degrees of freedom. If the calibration is accurate, there are slightly more: $(N*(N-1) - 3)*$(number of epochs), where the 3 is from the absolute position and intensity of the source, which are lost in self-calibration.

If the number of epochs is 1, and N is 11, the limit is 88 degrees of freedom in the source model. Based on trials with various model brightness distributions (discs, assemblages of points, asteroid shape models), the number of model parameters should be less than half the degrees of freedom if the image reconstruction is to be accurate to the diffraction limit. This limits the VLBA+GBT to 44 model parameters. Each synthesized beam on the target requires an independent model parameter, corresponding to its brightness. Therefore, the target cannot cover more than about 44 beams or, unless greatly elongated, an angular size less than about 6 times the synthesized beam. This means that as long as they are resolved, more distant objects are favored for image reconstruction. More distant objects have lower signal-to-noise. For a radar target to be an effective target for VLBA imaging, it must meet the constraints:

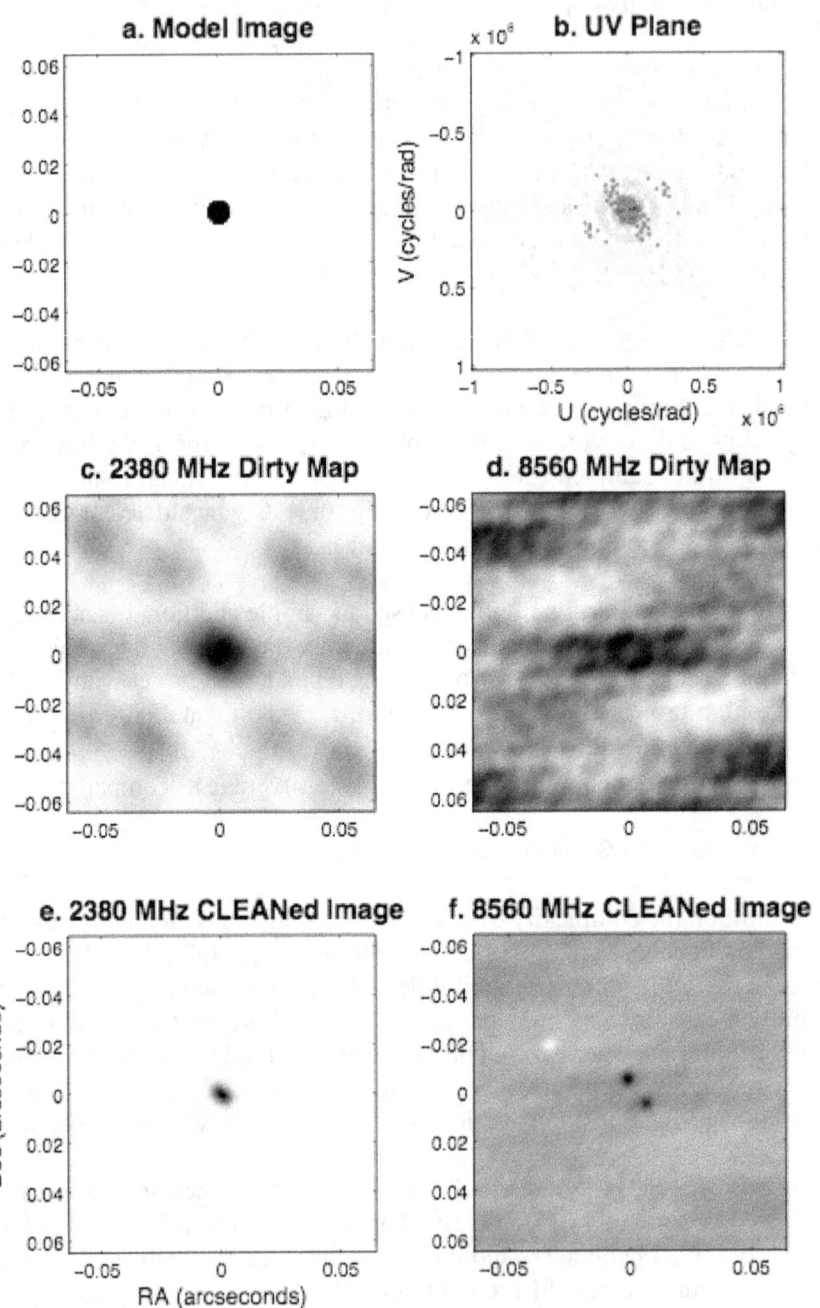

Fig. 4.4: a. Model image: a uniform disc. Grayscale is brightness (black high, white low). b. uv-plane and points for a VLBA snapshot at 2380 MHz (red) and 8560 MHz (green). Here grayscale is the amplitude of visibilities computed from the model image. U and V spatial frequencies are in units of cycles per radian of sky. c. Dirty map for 2380 MHz. d. Dirty map for 8560 MHz. e, f. Respective CLEAN restorations. In (e), the elongation of the source is due to the synthesized beam not being circular. In (f), the source is not recovered and the high- and low-brightness points are an artifact of CLEAN's response to over-resolved emission.

$$(4.6) \qquad \frac{2R}{d} \geq \frac{\lambda}{D_{max}} \qquad \text{- the target is resolved}$$

$$\text{and } \frac{2R}{d} \leq \frac{\approx (N/2)\lambda}{D_{max}} \qquad \text{- the target is not over-resolved}$$

$$\text{and } T_{noise} \leq \frac{G_T P_T}{d^2} a \left(\frac{\lambda}{D_{max}} \right)^2 \frac{\lambda^2}{2k} \qquad \text{- the target's radar echo can be detected}$$

where T_{noise} is the noise-equivalent-temperature of the receivers, and is ~30 K for the VLBA at 2380 MHz.

Eq. 4.6 excludes almost all radar targets from being the subjects of direct VLBA imaging. For example, the radar echoes from the objects 1999 AQ10, 1998 CS1, and 1994 CC were resolved and strong enough for radar-VLBI imaging during the first half of 2009. Only AQ10 would not have been over-resolved: it was ~6 VLBA+GBT 2380 MHz beams across while sufficiently strong for imaging, while CS1 and CC were both ~10 beams across.

If a radar echo is resolved in frequency, the number of degrees of freedom is increased: there are now N*(N-1)/2 baselines for each spectral channel for each epoch. If the spectral channels are very narrow compared to the bandwidth, each channel corresponds to a single line on the surface of the object – actually a strip with a width smaller than a beamwidth.

A linear source has one model parameter per beamwidth of length, plus two parameters determining its position on the sky and one its orientation. With self-calibration, each channel of visibility data therefore provides a maximum of ((N*(N-1) – 2N) – 3)/2 degrees of freedom to the image reconstruction, where the factor of two decrease comes from projecting the u-v plane to a single line. Assuming twice as many degrees of freedom as model parameters, this implies that the source cannot be wider than ((N*(N-1) – 2N) – 3)/4 beams. For the VLBA+GBT, this is ~21 beams. Tests with model linear sources agree with this number.

Fine spectral resolution largely fixes the over-resolution problem, at the expense of requiring still stronger targets as the echo power is spread out into at least as many bins as the echo spans beamwidths. Equation (4.6) changes to:

$$(4.7) \quad \frac{2R}{d} \geq \frac{\lambda}{D_{max}} \quad \text{and} \quad \frac{2R}{d} \leq \frac{\approx (N^2 - 3N)\lambda}{4D_{max}} \quad \text{and} \quad T_{noise} \leq \frac{G_T P_T}{2Rd} a \left(\frac{\lambda}{D_{max}} \right) \frac{\lambda^2}{2k}$$

While some targets are still over-resolved, this condition allows imaging of all three potential targets listed above, given sufficiently narrow spectral channels.

Only a subset of asteroid radar targets satisfies the conditions in Eqs. 4.6 and 4.7. Fig. 4.5 shows targets meeting the conditions, assuming VLBA+GBT receive, and approaching the Earth between 2009 and 2012. Due to Arecibo's lower transmit frequency and higher transmit power, there are many more known possible targets with Arecibo transmit than with Goldstone, 12 versus 2 for assumed rotation periods of 4 h.

Ten targets over 4 years is not sufficient to provide a representative survey of asteroid pole directions and spin states, but would be useful and much more capable than current adaptive optics.

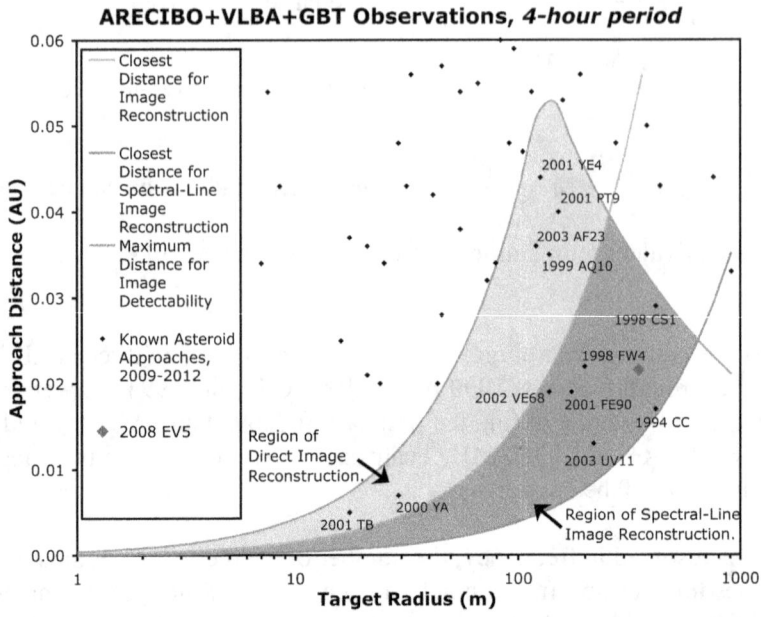

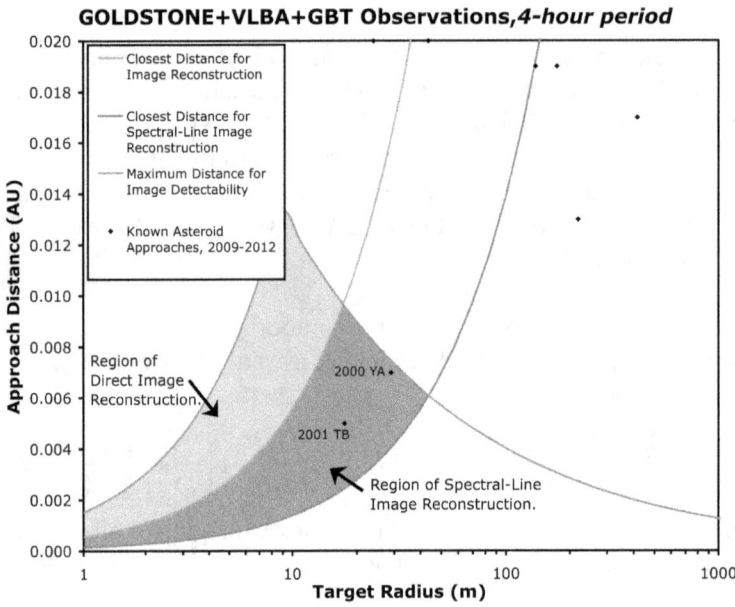

Fig. 4.5: Asteroid radar targets for 2009-2012 and the conditions given in Eqs. 4.6 and 4.7 for VLBA+GBT receive. Upper chart: Arecibo 12.6-cm 800 KW radar transmit. Lower chart: Goldstone 3.5-cm 450 KW transmit. This figure assumes radar albedo 0.1 and rotation period 4 h for all objects. The line specifying the maximum distance for imaging is set at small sizes by the asteroid being one beamwidth across and at large sizes by the echo having SNR/beam of 1. For objects rotating faster than 4 hours, the SNR/beam drops, while for objects rotating slower it rises. Note changed vertical scale.

Unfortunately, there is one more effect to be considered, and this makes radar interferometric imaging impossible for targets that are resolved – and therefore not useful for determining pole directions. This is the radar speckle pattern.

4.b.vi. Radar speckles and interferometry

On timescales of tens of seconds and longer, the radar echo flux received by an antenna is slowly varying and determined by the target object's size, shape, radar scattering properties, distance from Earth, and the gain variations of the antenna. The last could potentially be corrected, permitting image reconstruction. However, on short timescales (seconds or less), radar echoes vary randomly in brightness, from several times the average value to zero (Fig. 4.6). This is a radar speckle pattern (Green 1968, Kholin 1988, Margot et al. 2007).

Each point on the surface of the asteroid reflects the incident radar differently and acts as a radiator with random phase. Viewed from the Earth, the radiation from each pair of points interferes to produce a sinusoidal pattern of bright and dark speckles, as seen in Young's classic double-slit experiment. The points with the largest projected separation will produce the smallest speckles, with angular scale λ/d, where d is the diameter of the target and λ is the radar wavelength. The speckles from all possible pairs

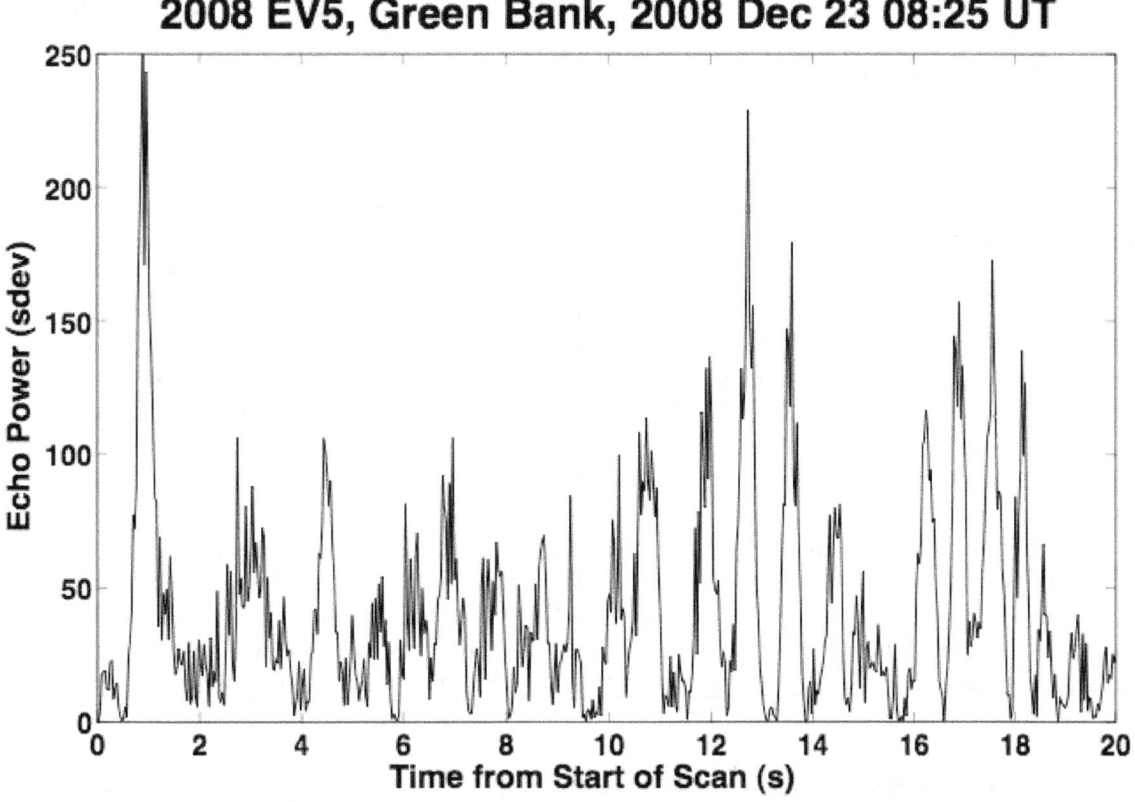

Figure 4.6. The echo power received by the Green Bank Telescope as a function of time, during radar observations of 2008 EV5. The point-to-point variations reflect the noise on the data, which is dominated by the self-noise of the echo. The dramatic ~0.5 s variations in echo power are radar speckles moving over the station. Over the 20-s span of the plot, the average echo power does not change significantly.

of points add together randomly to produce the speckle pattern received at the Earth, where the smallest speckles have length $L_{Speckle} = D_{target}\lambda/d$ (Fig. 4.7).

Stations further apart than the speckle scale receive signals that are not correlated with each other (Kholin 1988) – one station may be seeing a bright echo while a second is seeing nothing and a third sees intermediate brightness. If the target is resolved, each station receives a different $I(\sigma)$ and the assumptions of aperture synthesis as described above no longer hold. Averaging over many speckles to obtain the same $I(\sigma)$ takes longer than t_{coh}, making self-calibration impossible. This places the final and most important limit on radar interferometric imaging: the only targets that can be imaged are those that are not resolved.

There is at least one potential way around the speckle scale. Resolving the target in Doppler shift decreases its size in one dimension, and increases the speckle scale in that direction. If the array were ideally oriented (i.e., sufficient baselines all aligned in the long-speckle direction), this would provide the opportunity for self-calibration. In addition to requiring particular antenna arrangements that depend on a target's unknown spin state, this requires very strong radar targets that satisfy Eqs. 4.6 and 4.7. It is not likely to be useful.

Radar interferometry may be useful to obtain high-accuracy plane-of-sky astrometry, but it is not capable of determining pole directions. However, speckle patterns are themselves useful for this task.

4.c. Radar speckle patterns

4.c.i. Speckle patterns and pole directions

The speckle pattern is determined by the object's radar scattering properties and shape on all spatial scales larger than a fraction of a wavelength. In principle, sampling the entire pattern would permit a full reconstruction of the shape (Kholin 1988). That would require an implausibly large number of independent receiving stations (Sec. 4.c.ii). In practice, we are limited to measuring the speckles received at a small number of locations, insufficient to do more than measure $L_{Speckle}$. However, there is information in how the speckles change with time. As the asteroid rotates, the distance from the radar to each point on the surface changes, changing their relative phases and rotating the speckles in the same direction as the surface. On longer timescales (typically minutes), the speckles change in pattern as the incidence angle, and hence the radar scattering, at each point on the asteroid's surface changes.

Consider two stations that receive the same set of speckles from a radar target, either fortuitously aligned in the direction of speckle motion, or spaced much closer together than the speckle scale. They can be used to determine the sense of the asteroid's rotation by measuring the direction of speckle motion (Fig. 4.7, Busch et al. 2010a). Speckles from prograde-rotating asteroids will move from east to west, and those from retrograde rotators from west to east. Thus the motion of the speckle pattern indicates the sense of rotation.

The direction the speckles move can be determined by cross-correlating the echo power received at the two stations, to determine the relative time lag t_{lag} – the difference in arrival time of a speckle between the stations. The sign of the t_{lag} resolves prograde-retrograde ambiguities. The projected speckle velocity $|D/t_{lag}|$ is determined by the asteroid's rotation period P, the latitude θ of the sub-Earth

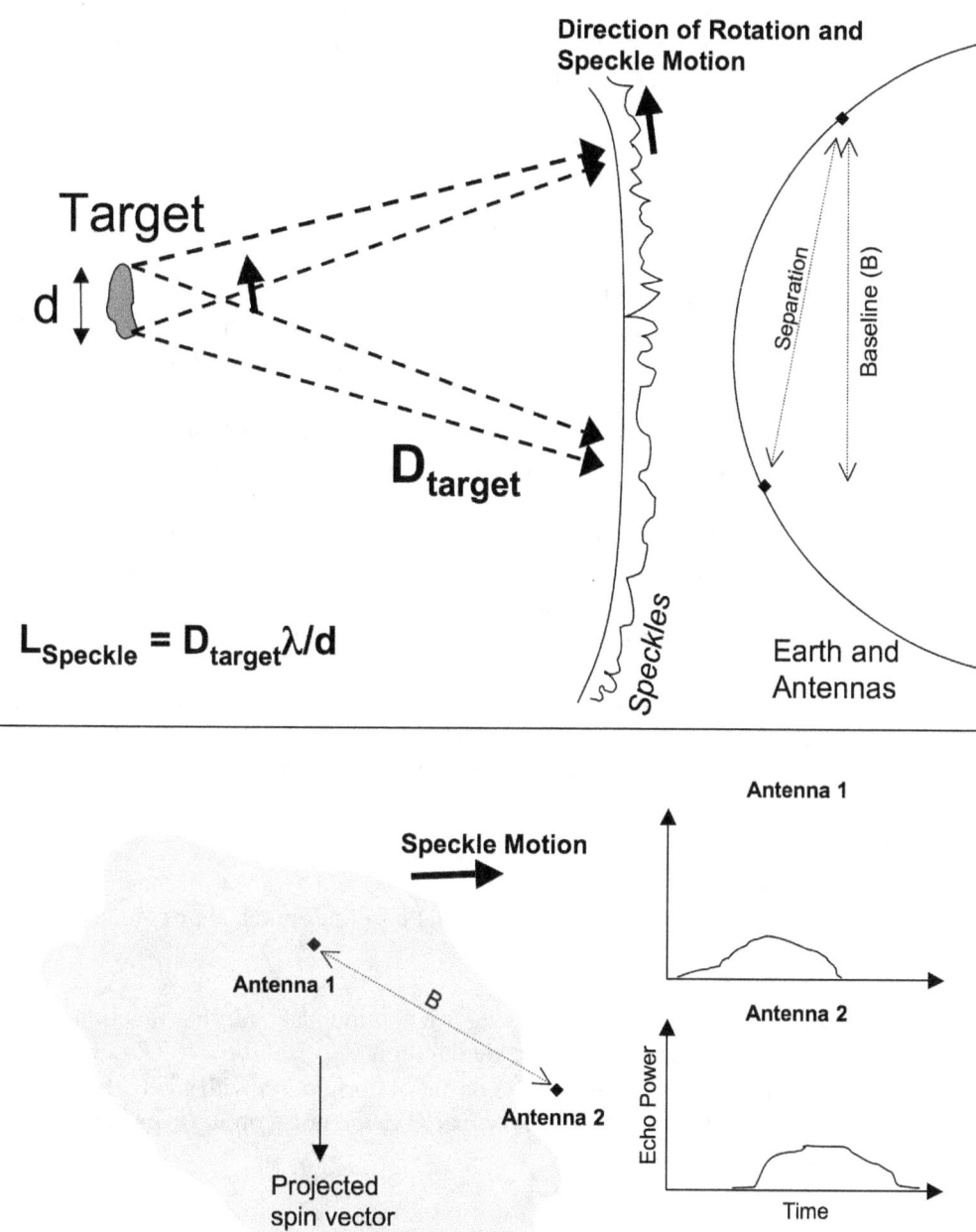

Fig. 4.7. Top. The reflected light from each point on the target's surface forms a wavefront and these interfere constructively or destructively, producing the random pattern of bright and dark speckles at the Earth. As the asteroid rotates, the phase at each point on the surface changes, moving the speckles in the same direction as the surface. Bottom. A single bright speckle passing over two antennas with baseline $B < L_{Speckle}$ and the echo power at each as a function of time. The average difference in arrival time over many speckles is equal to t_{lag}.

point on the asteroid, and the angle α between the spin vector and the antenna baseline (Kholin 1992, Margot et al. 2007). Because typical speckle velocities for asteroids are over 1000 km/s, the contributions from the Earth's rotation (<0.5 km/s) and the relative motion of the Earth and the target (always <90 km/s, usually ~10 km/s) can be ignored:

$$(4.8) \qquad \left| D / t_{lag} \right| \approx \frac{2\pi D_{target}\cos(\theta)}{P\sin(\alpha)}$$

For very close-approaching (low r) or slowly-rotating (high P) radar targets, Eq. 4.8 ceases to be a good approximation, and the Earth rotation and relative motion terms must be subtracted from the observed velocity to obtain the true speckle speed. For example, the asteroid 4179 Toutatis has $P > 100$ hours (Hudson & Ostro 1995), and will have speckle velocity ~130 km/s at $r = 0.046$ AU during its Earth approach in 2012, as compared to a contribution of 12 km/s from the asteroid-Earth velocity.

The magnitude of t_{lag} measures one dimension of the asteroid's pole direction (the angle α). Multiple baselines at appropriate separations, different angles and different times give the asteroid's complete spin vector, limited by the uncertainties in measuring the different t_{lag} values: the spin vector is the direction which gives the appropriate values of α for all the baselines. At least three baselines are required for a unique measurement of the spin vector by speckle tracking alone. These baselines can be between either the same or different pairs of stations, but must be at different times so that the target has moved and the sub-Earth latitude and projected spin axis have changed.

The maximum cross-correlation amplitude between two stations is determined by the radar echo's strength and the collecting area of the two receiving antennas, but most importantly by the baseline length projected along the direction of speckle motion. For $B\cos(\alpha) < L_{Speckle}$, where cross-correlation is possible, the relationship between the correlation amplitude and the projected baseline length is very roughly (Kholin 1992):

$$(4.9) \qquad C = e^{-2(B\cos(\alpha)/L_{Speckle})^2}$$

This very strong dependence on baseline length means that baselines much shorter than the speckle scale are required.

The cross-correlation as a function of time lag is a continuously varying function, with peaks with width in time approximately equal to the speckle duration $(L_{Speckle})/|B/t_{lag}| = \lambda P/(2\pi d)$. Smaller changes in time lag only partially shift speckles into or out of correlation with each other. Fortunately, given that the correlation amplitude is significant compared to the noise on it, t_{lag} can be determined to much less than the peak width.

The 1-σ uncertainty in t_{lag} is

$$(4.10) \qquad \Delta t_{lag} = \frac{1}{\sqrt{corSNR}} \frac{\lambda P}{2\pi d}$$

$$\text{where } corSNR = C \frac{P_{Eff}}{kT_{Eff}} b^{1/2} t_{Int}^{1/2}$$

$$\text{and} \qquad P_{Eff} = \frac{P_T G_T \sigma}{(4\pi)^2 r^4}\left(\frac{1}{A_1^2} + \frac{1}{A_2^2}\right)^{-1/2}$$

The correlation signal-to-noise ratio (*corSNR*) is determined by the correlation amplitude, the effective echo power (P_{Eff}), the echo bandwidth (*b*), the integration time (t_{Int}), and the effective temperature of the receivers (here assumed to be the same for both stations). The effective echo power is the geometric mean of the echo power received at the two stations, and is determined by the transmitter power (P_T) and gain (G_T), the distance to the target and its radar cross-section (σ), and the effective areas of the antennas (A_1 and A_2). The expression for P_{Eff} follows from the radar equation (Eq. 3.1).

The uncertainty in a time lag measurement cannot be converted blindly into the uncertainty in a measurement of α. Two nearby stations will see a series of speckles that are correlated with each other, but are not identical. For example, an elliptical speckle would both last longer and peak at a different time as viewed by a station that sees its center as compared to one offset to the side (Fig. 4.7). This introduces systematic errors into t_{lag}, which can be decreased by averaging and by using shorter baselines where the speckles are more similar to each other. In the limit of zero systematics, the uncertainty in α will be $\approx \Delta t_{lag} / t_{lag}$.

Δt_{lag} determines the minimum useful baseline length for speckle tracking: the distance for which t_{lag} becomes comparable to its uncertainty. There is therefore a range of suitable baselines:

$$D << D_{\text{target}} \lambda / d$$
$$(4.9) \quad D >> (2\pi D_{\text{target}} / P) \Delta t_{lag}$$

For objects with known pole direction, longer baselines can be used by observing at times when the baseline is aligned with the speckle motion (Margot et al. 2007). However, for new asteroid radar targets the use of baselines with $B > L_{Speckle}$ is a matter of luck, with the probability of cross-correlation being $<< r\lambda/dB$ for a random pole direction. Reliable speckle tracking requires baselines in the range given by Eq. 4.9.

4.c.ii. Structure of a speckle pattern

The radar speckle pattern of an asteroid is uniquely determined by the shape and radar scattering properties of its surface. However, in practice the pattern is indistinguishable from that of a randomly generated model surface. As a demonstration, consider the power spectrum of EV5's speckles as seen at Green Bank (Fig. 4.6) as compared to the power spectrum of a simulated series of speckles from a 450 m sphere with Lambertian scattering and random phase for each spot on the surface 12.6-cm across (Fig. 4.8). For all scales larger than the speckle scale, the power spectrum of the speckles is flat. Below that scale (which λ/d puts at 2.8e-4 radians), the spectrum falls off steeply to the noise limit of the receiver.

The fall off is due to aliasing: variations in the speckle pattern at higher spatial frequency than the speckle scale require two lower frequencies to interfere with each other and higher orders of aliasing necessarily contain less and less power. EV5's speckle power spectrum does start to fall off at slightly longer wavelengths than a 450-m Lambert model, even though a 450-m sphere is a fairly close approximation to its shape. This is consistent with the radar scattering law inferred from the delay-Doppler data (see Sec. 3.e.ii). Because of the low incidence angle, EV5's echo is stronger at the center of the disc and smaller separations and larger speckles contain somewhat more power.

On timescales longer than the speckle duration, the echo power measurements are not significantly correlated with each other and the long-wavelength spectrum is flat. The overall

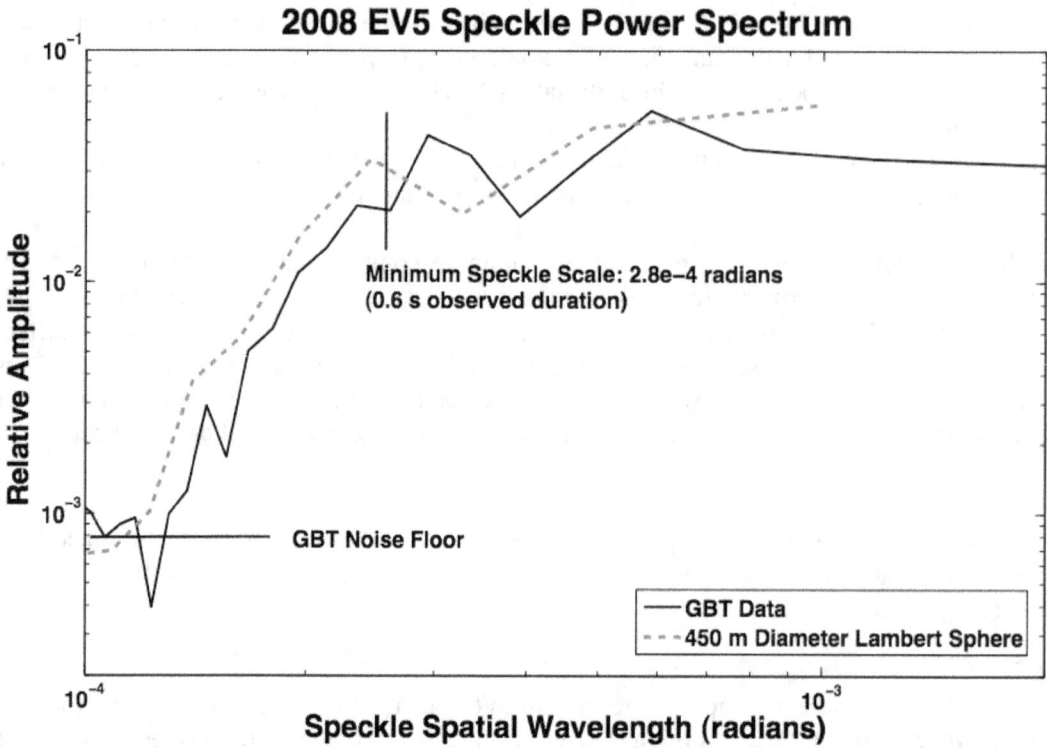

Figure 4.8: *Power spectrum of 2008 EV5 radar speckles as seen with Green Bank, and a simulated 450 m diameter sphere with Lambertian scattering and random phase. Speckle scale is measured in radians, obtained by specifying that the speckles move 2π radians in one 3.725 h rotation period. The flat spectrum for speckle scales larger than 2.8×10^{-4} radians is consistent with a surface rough on centimeter scales, the drop toward smaller spatial wavelengths is consistent with aliasing of large speckles, and the leveling off at speckle wavelengths less than about 10^{-4} is due to GBT's receiver noise. This spectrum contains 8 FFTs added together; point-to-point random variations are comparable to the amplitude.*

distribution of the measurements is well fit by a chi-square distribution with 1 degree of freedom. That distribution is expected for a surface composed of a large number of elements reflecting with uniform random phases (e.g. a rocky surface that is rough on scales larger than millimeters). Each speckle can be considered as a measurement reflecting the sum of the phases from each point on the surface along particular lines of sight. Considering points one wavelength apart, there are roughly 10^7 points on EV5. The vector addition of 10^7 randomly phased electric fields with any distribution of amplitudes produces a net field that is a very good approximation to Gaussian random, and echo power that is distributed in a chi-square fashion.

The information about EV5's shape and surface scattering is still encoded in the speckles. The speckle distribution tells us that each wavelength-scale spot on the visible surface is radiating independently. Each speckle can be considered an independent sum of the echoes from every one of those spots, added together with different phases determined by the topography of the surface on all scales larger than about 1 cm (when observed at Arecibo's 12.6-cm radar wavelength). In order to determine the phases uniquely, we would need a shape model of the asteroid accurate to 1 cm. Such a model would have $\gg 10^7$ parameters for EV5. To solve for such a model, we would need to have measured the strength of $\gg 10^7$ speckles. Since each speckle lasted 0.6 s, measuring $\gg 10^7$ speckles

56

would have equaled several telescope-months of time, which is impossible to arrange during a radar experiment.

For an elongated object, such as Itokawa or 1992 SK, the speckle scale will change as the object rotates and the maximum visible diameter changes. However, for all but the very smallest (few-meter) and fastest rotating (rotation periods of minutes) radar targets, it is impossible to obtain further shape information from the speckle pattern. For these very small targets, a few telescope-hours (and $\sim10^4$ speckles) might provide enough information for a shape model.

4.c.iii. Capabilities of speckle tracking

Speckle tracking is not particularly powerful in determining asteroid shapes. It is, however, very useful in determining pole directions. When combined with the spatial resolution of delay-Doppler, it enables unambiguous shape modeling.

If speckle tracking had been applied during the 2001 observations of 1950 DA, I would now know the pole direction and if there is indeed a potential Earth impact on March 16, 2880, rather than having to wait until 2032 for the next opportunity for detailed radar observations (Giorgini et al. 2002, Busch et al. 2007b). This is an example of the motivation for observing the speckle patterns of other strong radar targets.

Because stronger echoes provide better measurements of t_{lag}, the utility of speckle measurements is proportional to the sensitivity of the receiving antennas. With high sensitivity, speckle measurements of the spin state can be more accurate than measurements from imaging, as previously demonstrated for Mercury (Margot et al. 2007). For the strongest radar targets and the most sensitive antenna pair currently available (Jodrell Bank –Efflesberg), speckle tracking can measure the pole direction to better than 1°; there are about two such targets each year. With sub-degree measurements, it may be possible to directly track small changes in asteroid pole directions; produced by YORP, collisions, shape changes, or tidal interactions in binary systems – such as the precession of the spin axis of 1999 KW4 Alpha (Ostro et al. 2006, Scheeres et al. 2006). For objects in non-principal axis rotation states (wobbling or tumbling with a fixed angular momentum vector but no fixed pole direction), accurate measurements of the spin state – on timescales of days to years – would provide the moments of inertia and information about the internal mass distribution (Hudson & Ostro 1995).

Of all of the techniques I studied, radar speckle tracking is the best suited to measure the spin states of near-Earth asteroids, although, again, it is most useful in combination with delay-Doppler data. Therefore, I have developed software to use the stations of the VLBA, and potentially other radio telescopes, to track speckles. I have also used this code to resolve the ambiguity in the pole direction of the asteroid 2008 EV5 that I described in Chap. 3. I describe the code and the observations in Chap. 5.

5. Implementing Radar Speckle Tracking

Conceptually, radar speckle tracking is very similar to radio interferometry: both rely on cross-correlating the voltages received by radio receivers at stations separated by significant distances. Speckle tracking differs from interferometry in two respects. Rather than synching the signals in time and phase, I deliberately offset them to determine the relative time lag of the speckles; and the phase of cross-correlation is not relevant. More important for implementing radar speckle tracking is the narrow bandwidth of radar echoes: typical bandwidths at S-band are of order 1 Hz. The hardware-optimized correlators used by most radio interferometers (e.g. Napier et al. 1994) are not able to resolve such narrow bandwidths. Therefore, for my radar speckle observations with the VLBA, I needed to develop a special-purpose correlator program.

5.a. Radar speckle processing for the Very Long Baseline Array

5.a.i. Computation of cross-correlation

In radio interferometry, the visibility is computed as the cross-correlation between two voltage streams for each frequency (Sec. 4). The cross-correlation computations for radar speckle tracking are similar to those to compute visibilities, with the modifications of the time lag between the stations and the cross-correlation using the echo power received at each station rather than the voltages (amplitude rather than amplitude and phase).

To compute the cross-correlation as a function of t_{lag} for the echo requires correlating the power in the echo's frequency band received at two stations for each candidate time lag. It is possible to compute the correlation first and then apply the Fourier transform to separate out the frequency channels, the 'XF' design used in many hardware correlators. However, for correlation in software it is computationally simpler to apply the Fast Fourier Transform to a block of data from each antenna first, since the Fourier transform of the visibility is $F(V_{ij}) = F(V_i)F(V_j)$ - that is, correlation in the time domain becomes a simple multiplication in the frequency domain. Applying the FFT before the correlation is termed an 'FX' correlator.

Any correlator must perform a very large number of FFTs, which are most of the computational burden. Due to the requirements of computing the correlation of a broadband signal, most radio interferometers use specially built hardware correlators with pre-determined minimum channel bandwidth. For example, the VLBA Mk V correlator had a minimum bandwidth of 120 Hz (Benson 1995). A correlator channel bandwidth far larger than the echo bandwidth yields a greatly decreased signal-to-noise. This motivates correlation in software, where the minimum channel width is limited only by the runtime, the accuracy of the frequency calibration, and atmospheric stability. While several software correlators already exist (e.g. Black et al. 2005), the most relevant program for this project is the DiFX (DIstributed FX) correlator program, originally written by Adam Deller (Deller et al. 2007).

A version of DiFX for the VLBA has been developed at the NRAO. While it is now in general use, it was not yet available when I began this project. Even now, runtime considerations on the Array Operations Center computer cluster limit the spectral resolution of VLBA observations DiFX to about 7 Hz. This limitation will be relaxed with the next version of DiFX, to be available in mid-2010. Therefore, I wrote my own correlator program to process asteroid radar VLBA observations.

5.a.ii. The software correlator

My software correlator has been designed strictly for narrow-band observations. Since the maximum relative Doppler shifts of a radar signal between stations on the Earth is <<100 kHz, the correlator assumes that the input data has been downsampled to no more than several hundred thousand samples a second (source code is provided in Appendix 2). Because I have not implemented any multi-processor capability, the correlator also runs at far less than real time.

To run VLBA data through the correlator, I use pre-existing routines to extract and downsample raw time-stamped data from VLBA Mark 5 data module files that have been copied to archival disc. The main correlator program reads in the data and their timestamps and synchs the voltage streams from all stations in time to the nearest half-sample. It then computes the near-field corrections for each station from an external ephemeris file from the JPL Horizons system (http://ssd.jpl.nasa.gov/?horizons). The delay corrections are applied by further shifting of the data streams. The Doppler shift corrections can either be turned off, for diagnostic output or checking the accuracy of the ephemeris, or applied to each data stream to move the echo to a very low but non-zero nominal frequency. The data are then filtered to a narrow bandwidth and downsampled further prior to the cross-correlation.

After filtering and downsampling, the correlator computes the cross-correlation for each pair of stations for a block of data of specific length, to provide the desired frequency resolution, and a number of successive blocks are added together. Output of the correlator for auto-correlations (each station correlated with itself at zero time lag), using data from the December 2008 VLBA+GBT observations of the 2008 EV5, is shown in Fig. 5.1. The program accurately unpacks the data, applies the near-field corrections, and can achieve frequency resolution of better than 0.5 Hz.

The software correlator's theoretical frequency resolution can be improved indefinitely by FFTing larger blocks of data - either accepting increased runtime or implementing a prefilter routine to decrease the data rate even further. The fundamental limit on frequency resolution is the phase stability of the atmosphere. At 2380 MHz, the atmospheric phase delay is stable to within one radian for roughly 100 s, with considerable variations due to weather, so that frequency resolution less than 0.01 Hz is problematic. In order to approach this limit, the Doppler shift at each station must be taken out before the FFT is computed, by phase-rotating the individual samples. Doppler shifts due to Earth's rotation change by ~0.1 Hz/s. Any frequency resolution less than 0.5 Hz will result in the radar echo smearing across frequency bins during one FFT, unless the Doppler shifts are taken out first. With 0.01 Hz resolution, the echoes of all but the very slowest-rotating asteroids can be resolved.

If the averaging time – the interval over which a cross-correlation is computed – is much longer than the length of one FFT, then the noise on the cross-correlation will be normally distributed. While broadband sources have very short FFT lengths, asteroid radar echoes do not. For example, a frequency resolution of 0.5 Hz requires FFTs that are each 2 s long. If the averaging time is not much greater than this, the variance of the computed average will be increased, following a chi-square rather than a normal distribution (Ostro 1993). Fortunately, for speckle tracking applications, the cross-correlations are computed over several minutes or more, to maximize signal-to-noise, and thus the noise is approximately Gaussian.

As a further check, I processed the data both with my narrow-band software correlator and with DiFX (Deller et al. 2007). For EV5, the minimum integration time was 0.325 s. In order to well-

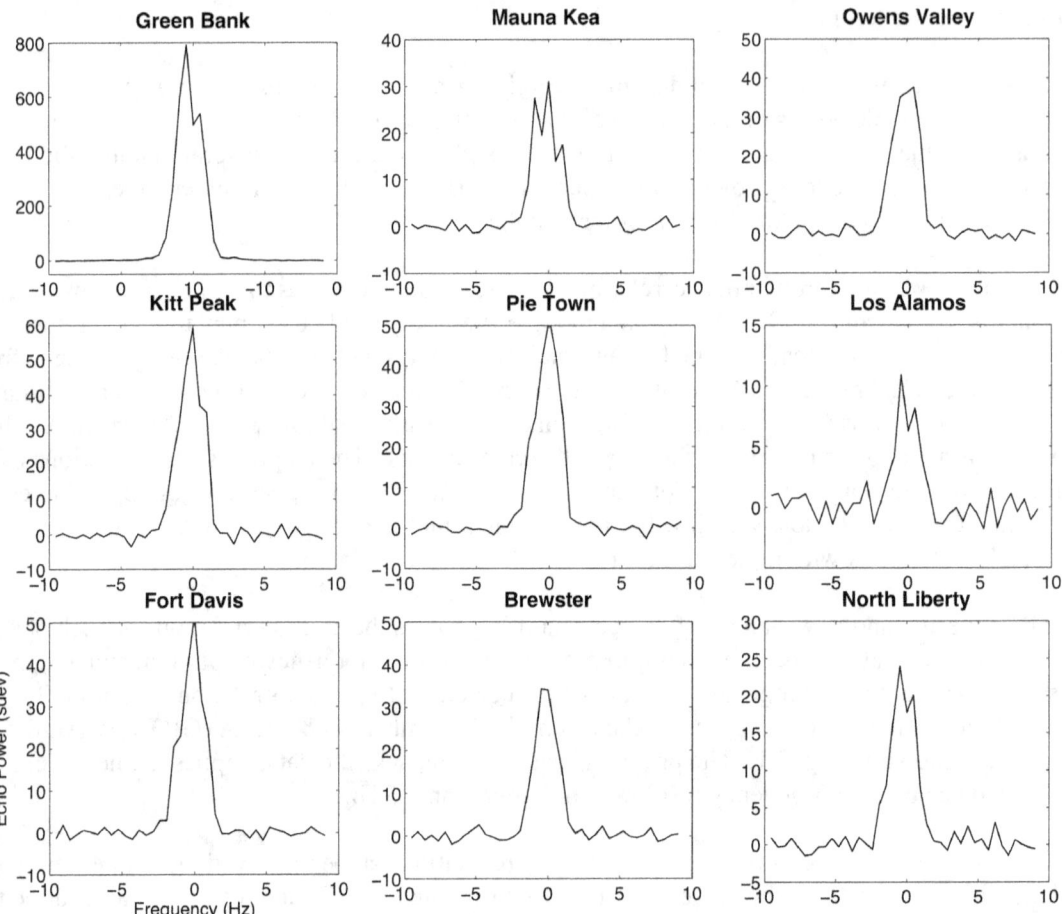

Fig. 5.1: Radar echo of 2008 EV5 as received by Green Bank and 8 VLBA stations on 2008 Dec 23 and processed with my software correlator. Frequency is given relative to the 2380 MHz at geocenter. Resolution 0.48 Hz; the echo is ~3.5 Hz wide. These spectra are 120-s integrations (2008 Dec 23 08:25-08:27 UT) and average over many speckles. Variations in echo strength correspond to variations in antenna gain; Green Bank's higher gain compared to the 25-m VLBA antennas translates to higher signal to noise.

resolve the speckles and to maximize the signal-to-noise ratio of the cross-correlations, I required the fine time resolution of my narrow-band correlator.

The two programs produce consistent output (Fig. 5.2). Downsampling the narrow-band output to the same time resolution produces a time series of echo power that matches that produced by DiFX to the level expected due to inevitable differences in the data binning. Future upgrades to DiFX should permit shorter integration times and narrower channels and make speckle observations a standard VLBA mode.

5.a.iii. Changing t_{lag}

The correlator cross-correlates all of the data files it is supplied with each other, for each of a user-specified set of time lags, and the processing time scales linearly with the number of different time lags it checks. To be sure of locating the time lag with maximum cross-correlation (the true travel time of the speckles), the correlator must search at least potential lags from $-2\pi PD/D_{target}$ to $2\pi PD/D_{target}$. These two lags correspond to the speckles moving straight along the baseline in one direction and to speckles moving directly in the opposite direction.

5.a.iv. Station selection for speckle observations

The 25-m VLBA stations are not ideal for speckle tracking. They can provide rotation information for perhaps the 10 strongest radar targets each year (Fig. 5.3). There are larger telescopes available, but only a few with suitable baselines. These include Arecibo, Goldstone, the Owens Valley Radio Observatory, Jodrell Bank, and Efflesberg (Table 5.1, Hjellming 2000). For small radar targets, with speckle scales measured in the thousands of kilometers, there are other potential stations. For Goldstone transmit and large speckles, there are options for stations in Japan and Australia. These are tentative: they require Goldstone transmit, and would be useful only for the smallest targets, such as a 3-m object at 1 lunar distance. Targets that small will not be resolved by delay-Doppler imaging, but there might be interesting science in determining their pole directions. Such small objects are presumably single blocks and, depending on their rotation rates, should be at the transition point where thermal conduction begins to cancel out the effects of YORP and Yarkovsky.

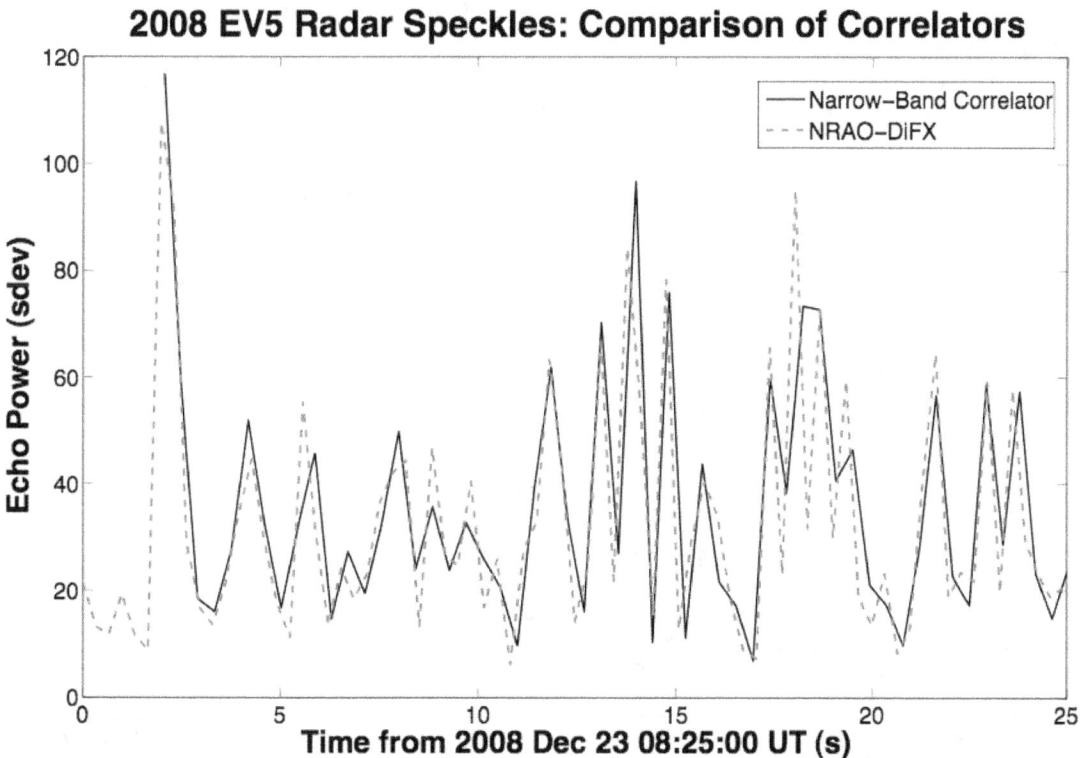

Fig. 5.2: The 2008 EV5 data from GBT on 2008 Dec 23 presented in Fig. 4.6 processed with time resolution of 0.325 s, using both my narrow-band correlator and DiFX. At this resolution, the speckles are only slightly resolved. The mismatches in amplitude are due to slight differences in the binning of the data.

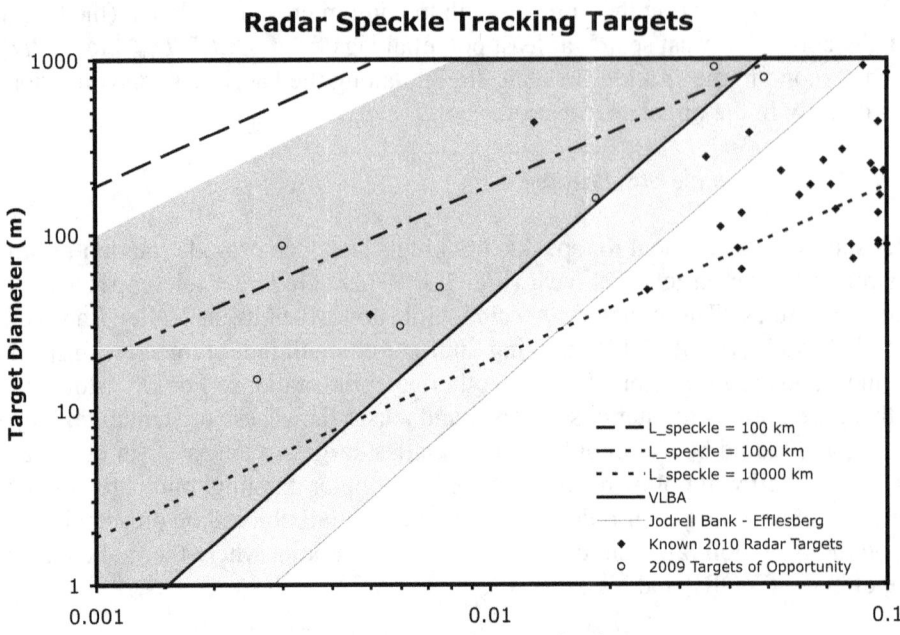

Fig. 5.3. *Asteroid radar targets and the range of approach distance and diameter suitable for speckle tracking with the VLBA (dark shaded region) and with the Jodrell Bank and Efflesberg telescopes (dark and light shaded regions). There are ~10 speckle targets each year, the majority of which are targets of opportunity <= 100 m across.*

Table 5.1: Antenna Pairs for Speckle Tracking

Antennas				Baseline (km)	Sensitivity /VLBA
Goldstone 70-m	USA	DSS-24,25,26 34-m	USA	10	2.8
Goldstone 70-m	USA	DSS-13 34-m	USA	22	2.8
Usuda 64-m	Japan	Nobeyama 45-m	Japan	28	3.3
Madrid 70-m	Spain	Yebes 40-m	Spain	100	3.0
Jodrell Bank 76-m	UK	Cambridge 32-m	UK	198	2.6
Goldstone 70-m	USA	Owens Valley 40-m	USA	233	3.1
Canberra 70-m	Aus	Parkes 64-m	Aus	274	6.3
Effelsberg 100-m	Ger	Cambridge 32-m	UK	510	2.6
Haystack 46-m	USA	Algonquin 45-m	Canada	642	2.5
Effelsberg 100-m	Ger	Jodrell Bank 76-m	UK	700	11.7
Green Bank 100-m	USA	Haystack 46-m	UK	846	3.1
Green Bank 100-m	USA	Algonquin 45-m	Canada	847	4.0

Combinations of antennas for radar speckle tracking sorted by baseline length. The baselines given here have been rounded to the nearest 1 km and are the straight-line separations between the antennas. During an observation, the baseline length decreases depending on the direction to the target. Sensitivities are given as ratios of (speckle correlation SNR for antenna pair)/(speckle correlation SNR for two VLBA stations) for the same target and are approximate. Only baselines less than 1000 km are listed here. Shorter or longer baselines will be preferred for any given target (see text). DSS refers to antennas at Deep Space Network Sites, located at the Goldstone Deep Space Communications Complex. Data from telescope parameters given in Hjellming (2000).

While planning observations of new radar targets, appropriate baselines must be identified. However, diameter estimates from optical photometry can differ from the true size by up to a factor of two due to variations in albedo and the maximum and minimum useful baseline length will be uncertain by that factor as well. For most strong radar targets, this leaves a range of useful baselines between tens and hundreds of kilometers. The larger antennas provide a few more targets, but more importantly provide much higher signal-to-noise ratio and therefore more accurate pole directions.

5.b. Application to a near-Earth asteroid: 2008 EV5

As I have described (Sec. 3.e.ii), the VLBA and GBT observed EV5 during December 2008. Since EV5 has a diameter of 450 m, on 2008 Dec 23 the speckle scale at 12.6 cm was ~900 km. This is consistent with the speckle duration of 0.65 ± 0.05 s, as observed at Green Bank (Fig. 4.6).

The closest pairs of antennas in the VLBA are the stations in the southwest United State: Pie Town and Los Alamos, New Mexico; Kitt Peak, Arizona; and Fort Davis, Texas. For speckle cross correlation, we considered the closest VLBA antenna pairs: Pie Town (PT) and Los Alamos (LA), New Mexico; Kitt Peak (KP), Arizona; and Fort Davis (FD), Texas. The shortest baseline during our EV5 observations was Pie Town to Los Alamos (PT-LA): 235 km. The highest cross-correlation was obtained at PT-LA t_{lag} = -0.17 $\pm$ 0.05 s. The speckles moved from west to east, arriving at Pie Town first (Fig. 5.5). Therefore, EV5 has a *retrograde* pole direction.

The speckle velocity along PT-LA was $|B/t_{lag}|$ = 1380 (+ 700 -500) km/s, implying that the baseline was angled at 90° (+ 0° -40°) relative to EV5's rotation axis (90° corresponds to 1350 km/s). This is consistent with the retrograde pole direction estimated by inverting the delay-Doppler data (Busch et al. 2010).

The second- and third-shortest VLBA baselines (KP-PT and PT-FD) also show peaks in cross-correlation, but they are weaker and have correspondingly larger Δt_{lag} (Fig. 5.4). This is expected from Eq. 3. For $B\cos(\alpha)$ = 450 km = ($L_{Speckle}/2$), the cross-correlation amplitude between stations should drop by ~40% as compared to a baseline of nearly zero length. Because the KP-PT and PT-FD baselines were between 400 and   500 km long, they have weaker cross-correlations than PT-LA does. All the other baselines in the array (LA-FD and longer) do not have significant cross-correlation.

The PT-FD cross-correlation has significantly higher signal-to-noise-ratio than that for KP-PT, despite the somewhat larger B and lower correlation amplitude. This reflects more accurate pointing (and hence higher gain and greater effective area) at Fort Davis. The time lags along the KP-PT and PT-FD baselines are consistent with the delay-Doppler pole direction. The KP-PT baseline had $\alpha \approx 10°$, producing very small t_{lag}.

The uncertainties in our measurements of t_{lag} are conservative and include an upper limit of the systematic error from differences between the speckles seen by the two stations. The systematic uncertainty in t_{lag} as measured by a single speckle is comparable to or less than the speckle duration, and will average down as the square root of the number of speckles. Because the cross correlations each included ~250 speckles with duration ~0.65 s, the systematic error will be <~0.042 s. Combining this with the formal Δt_{lag} = 0.018 s gives an overall uncertainty of <= 0.05 s.

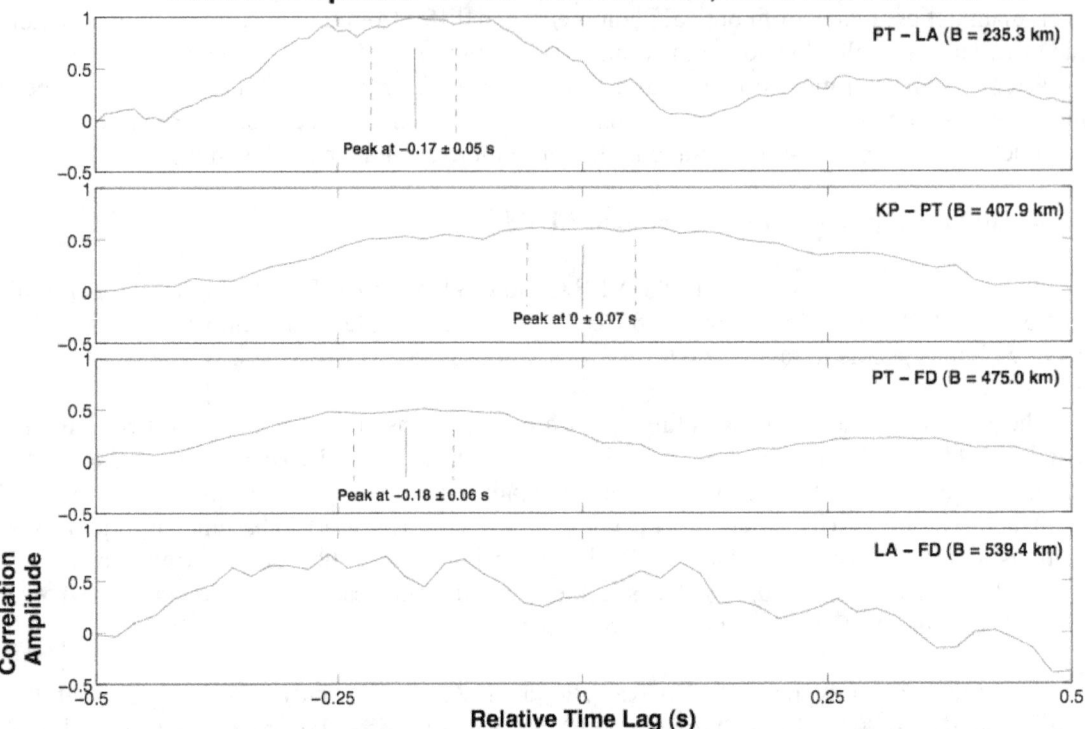

Fig. 5.4. Cross-correlation of the echo power received at the VLBA stations as a function of relative time lag for the 4 shortest baselines in the array (labels give stations and baseline lengths). t_{lag} was sampled at 0.01 s intervals. The peaks at negative t_{lag} indicate retrograde rotation, and are consistent with EV5's rotation period and the angles between the spin axis and the baselines (see text). The uncertainties in our estimates of the position of the peaks include upper limits on systematic errors. For LA-FD, the correlation peak is too weak for a fit to be meaningful.

Regardless, I have tracked the motion of EV5's radar speckle pattern along three baselines, and used this information to split the delay-Doppler ambiguity in its spin state (Fig. 5.5). Now that the narrow-band software correlator is available, future speckle tracking observations should be routine. I am currently planning to make speckle tracking a regular part of the observing schedule for strong Arecibo radar targets. The next two known speckle targets are 2005 YU55 in April 2010 and 2003 UV11 in October 2010. There will also be newly discovered objects with sufficiently strong radar echoes, at the rate of 3-10 per year.

5.c. Other applications of speckle tracking

As a final note, in addition to asteroids, there are planetary applications of speckle tracking. Speckle tracking has already been used to measure Mercury's libration, demonstrating the presence of a liquid outer core (Kholin 2004, Margot et al. 2007). The speckles from a planet are much smaller than those from an asteroid: ~1 km. Since the pole directions of the planets are known, the motion of the speckles across the Earth can be predicted, and times chosen when they move parallel to a particular baseline – so called 'magic moments' (Margot et al. 2007).

64

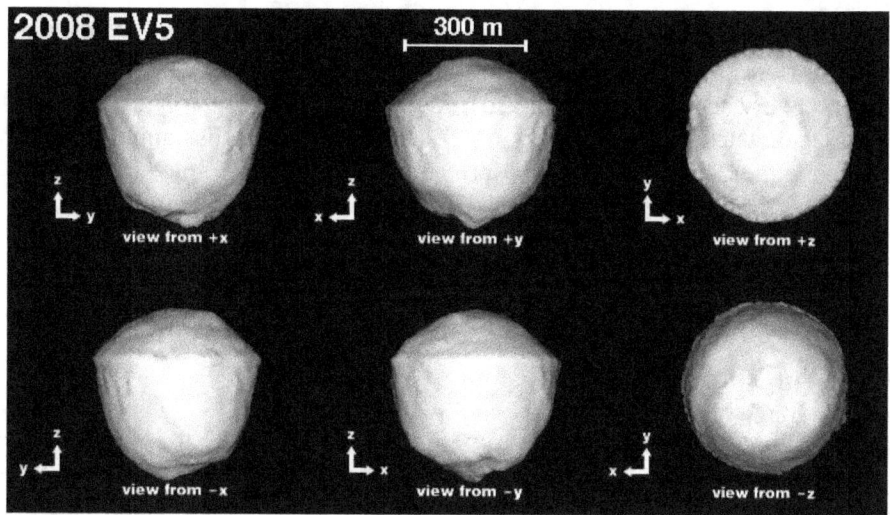

Fig. 5.5. *Principal axis views of the best current EV5 shape model. The yellow shading denotes areas that were not visible at incidence angles less than 60°. The exact latitude of the ridge and position of the accompanying depression are uncertain; the ridge may lie at latitudes between 30° N (+z) and 30° S (-z). Figure from Busch et al. 2010b.*

For Mars and the Moon, spacecraft telemetry provides accurate rotation information, making speckle tracking unnecessary. I am, however, interested in the possibility of tracking the speckle pattern of Venus and the Galilean satellites of Jupiter, to search for variations in the length-of-day or detectable polar wander. This would rely on long baselines and very large antennas (GBT, Arecibo, Goldstone). Similarly, for asteroids with known pole directions, speckle tracking with long baselines during magic moments would provide greatly refined measurements of their spin states – perhaps sufficient to measure precession and moments of inertia, and infer the internal mass distribution of the object. As an example, this would illuminate the size distribution of the components of a rubble pile.

6. Future Possibilities in Asteroid Science

Asteroid discovery and dynamics and asteroid morphology are rapidly evolving fields, and any forecasts I make are guaranteed to be incorrect in the details. With that disclaimer, I see three major future developments of the techniques I have described to study asteroids' physical properties. For the near-Earth asteroids, as the next generation of surveys becomes active, the number of radar targets will dramatically increase. Both speckle tracking and improved delay-Doppler observations will be essential to keep pace with the discoveries. For the main belt, submillimeter observations with the ALMA array will soon make interferometry an attractive prospect.

6.a. Effect of Pan-STARRS, LSST, and other surveys

Currently, near-Earth asteroid survey programs have located ~90% of objects larger than 1 km, but only 1%-10% of 100 m objects, based on the rate of discovery and the brightness distribution of known objects (Harris 2008). The next generation of survey telescopes, particularly the Pan-STARRS (Jedicke et al. 2007) and LSST projects (Ivezic et al. 2007, LSST 2010), plan to locate the majority of objects in the 100-m size range within the next 10-20 years.

If and when the roughly 100,000 near-Earth objects larger than 100 m are discovered, there will be an increase in the number of asteroid radar targets by roughly an order of magnitude. Rather than perhaps 10 targets per year suitable for high-resolution shape modeling, there will be hundreds, with thousands of weaker targets suitable for radar astrometry only. The latter will include many potential Earth impactors. Cases like 1950 DA and 99942 Apophis (Giorgini et al. 2008) will become commonplace. There will be additional fireball objects like 2008 TC3, discovered only days before impact.

Many of the new radar imaging targets will be very small (~10 m), and discovered only a few weeks before making a close Earth approach. Observing them with delay-Doppler will require new radar transmitters with finer delay resolution (Slade et al. 2009) and a significant increase in the fraction of Arecibo and Goldstone time dedicated to radar observations (Giorgini et al. 2009). Similarly, speckle observations will require a significant block of target-of-opportunity time on suitable telescopes. If both the necessary observing time and the capability to construct physical models are available, the representative sample of near-Earth objects I described in Chap. 2 (shapes, spin states, and composition information for several hundred objects) will be available within a few years of the start of an LSST-scale survey.

LSST or a comparable survey will also provide lightcurves for the majority of the objects discovered: covering the entire sky once per night produces sparsely-sampled lightcurves that can be inverted to produce period estimates and approximate pole directions (Durech et al. 2007). As with the lightcurves currently available, these will be most useful when combined with radar data (Sec. 4.a).

6.b. Future radar facilities

I have assumed that both of the current planetary radars will be available indefinitely. For at least the next decade, this is probably correct. However, in the longer term both Arecibo and the Goldstone 70-m antenna may be decommissioned. In particular, the Goldstone 70-m may be decommissioned when it is no longer needed to ensure contact with the Voyager and Cassini spacecraft and high-bandwidth communication with new missions can be accomplished with smaller antennas (MacNeal et al. 2007). A new planetary radar will then be needed, and it must be capable of observing a large number of targets.

Several spacecraft now carry transmitters and receivers at Ka-band – 32 GHz or 0.94 cm – for example, the Mars Reconnaissance Orbiter (Shambayati et al. 2007). There are two advantages to radar observations at this frequency. The gain of the transmitter and receiver are both higher by a factor of 14 than for the same antennas at 3.5 cm (in accordance with Eq. 3.1). Given 500 KW transmit power and equal receiver noise, one of the 34-m Deep Space Network antennas at 0.95 cm would be approximately 45 times more sensitive than the Goldstone radar is at X-band and 4 times more sensitive than Arecibo is at S-band. Sensitivity would be further improved by transmitting and receiving with larger antennas, such as a duplicate of Green Bank or potentially the largest single elements of the Square Kilometer Array (Ostro 1997). Inevitably, there will be a compromise between increased sensitivity and lower transmitter power (and hence lower operating cost).

The second advantage to a higher carrier frequency is the larger amount of spectrum available. Modulating the carrier wave with delay coding inevitably broadens the transmitted bandwidth by the coding rate. The 0.05 µs or 20 MHz limit on the delay coding at Arecibo is in part artificial: the radar has only been allocated a narrow band, so that the transmitter sidelobes do not cause interference to communications. At Goldstone, the new high-resolution receiver will be similarly limited. At Ka-band, bandwidth allocations are hundreds of megahertz wide (Office of Spectrum Management 2003), and the resolution of delay-Doppler images could be improved to perhaps 1 m (coding at 6.7 ns or 150 MHz). Constructing a 32-GHz radar would require significant engineering, but much of the technical development will follow from programs to extend spacecraft communications to Ka-band.

There are some disadvantages to radar observations at such high frequencies. Absorption due to water vapor becomes high at low angles above the horizon, limiting sky coverage and observing intervals. Also, radar speckle scales at Ka-band will be factors of 13-14 smaller than for 12.6 cm transmit. This dramatically limits the antenna baselines that can be used, for equal-sized targets. Rather than VLBI, speckle tracking at Ka-band would use local arrays such as the Very Large Array, and the precision of pole direction estimates would decrease. Given the atmospheric stability timescale between 20 and 40 GHz (Carilli & Holdaway 1999), interferometry is still not possible at Ka-band. Asteroid radar astronomy would greatly benefit from a few-GHz frequency radar being maintained beyond the lifespan of Arecibo or Goldstone.

6.c. ALMA, CARMA, and Thermal Radiation

Interferometry will become a powerful tool to study asteroids, but only by going to still higher frequencies. The Atacama Large Millimeter Array (ALMA) is currently under construction at the Llano de Chajnantor site in Chile. When complete in 2012, it is to contain fifty 12-m antennas, as well as smaller dishes to fill in short antenna separations. ALMA will be the most sensitive 100 – 960 GHZ (3 mm – 0.3 mm) observatory ever constructed (ALMA 2004). Because ALMA observes at such high frequencies, it can detect asteroids without radar illumination, using the thermal emission from the near subsurface.

Previous discussions of applications of ALMA's capabilities to planetary science and asteroids in particular (Gurwell 2004, Lovell 2007) have focused on detecting thermal emission and constructing models of the targets based on a series of total-intensity measurements, a thermal lightcurve (e.g. Chamberlin et al. 2007). I observed (Busch 2009) that ALMA's spatial resolution will allow high-resolution imaging as well as lightcurve measurements. In its largest configuration, the array will have baselines of 14 km. At the high frequency end, 790-950 GHz, this corresponds to resolution of 5 mas or 4 km at 1 AU. Given ALMA's sensitivity limits, this will provide the potential for rapid mapping of the shapes and surface temperature distributions of all asteroids that span at least a few times the resolution. This implies imaging of objects larger than about 40 km across in the main belt (~700 objects), and larger than about 75 km in the Jupiter Trojan population (~100 objects). While speckle tracking and delay-Doppler imaging can survey the near-Earth asteroids, ALMA will be able to map most of the mass of the main belt.

Such an ALMA survey would be invaluable. As with the near-Earth population, many fundamental questions about the main asteroid belt remain unanswered. The composition of perhaps half of the 700 largest objects is unknown, in several dozen cases despite optical and near-infrared spectroscopy and radar albedo measurements (e.g. Rivkin et al. 2000; Shepard et al. 2008). While we now understand the importance of thermal emission on asteroids' orbits and spin states via Yarkovsky and YORP, in almost all cases we cannot model these effects except in a statistical sense. Most modeling of Yarkovsky and YORP has considered populations of objects, assuming spherical or ellipsoidal shapes. As I described in Chaps. 2 & 3, only models of particular objects with known shapes, such as 1950 DA or Golevka (Chesley et al. 2003), produce accurate trajectory predictions.

Accurate models of Yarkovsky and YORP require knowing an asteroid's surface temperature distribution as well as its shape. As I have shown, radar observations are limited in range and target size to the near-Earths and the very largest main-belt objects, and do not map the temperature distribution directly. All of this motivates imaging asteroids with high spatial resolution in the sub-millimeter range, where their thermal emission can be detected – that is, observing them with ALMA.

Applying the observational limits on source size that I derived for interferometry (Eq. 4.6) for ALMA with 50 stations, the number of resolution elements covered by a source for a snapshot observation can be as high as 1175 and therefore the source should not be larger than about 35 times the resolution of the array. With the exceptions of Ceres, Pallas, and Vesta, no main-belt asteroid exceeds this angular size. Because these very largest objects can be observed when the array is in a more compact configuration, ALMA will not over-resolve asteroid targets.

For ALMA, as for any interferometer, imaging ability changes somewhat depending on the structure of the source and the orientation of the array. Therefore, I have simulated an ALMA 750 GHz (0.4 mm) continuum image of the main-belt asteroid 216 Kleopatra, whose shape has been modeled from radar observations (Ostro et al. 2000) and validated by adaptive optics images (Marchis et al. 2008). My simulation includes only Kleopatra, and not its two small satellites (5 and 3 km across), which would be marginally detectable, but not resolved, by ALMA. The results of the simulation are shown in Fig. 6.1. I took station positions from the ALMA strawman wide configuration model (ALMA 2004), produced model visibilities, and applied CLEAN to deconvolve the resulting dirty map. The reconstruction is accurate to the diffraction limit of the array.

Beyond resolution and image reconstruction, ALMA's sensitivity will limit imaging. The main belt asteroids have surface temperatures of 120-240 K. At 750 GHz, ALMA's designed continuum sensitivity for a 60 s integration and 6 mas resolution is 72 K (ALMA 2006). For an object at 2 AU from

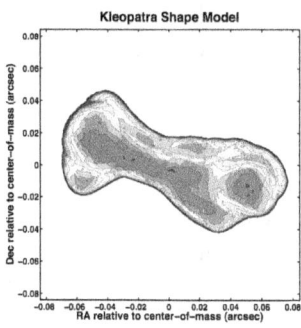
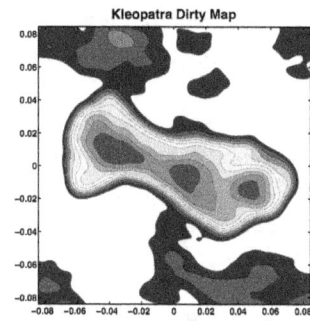
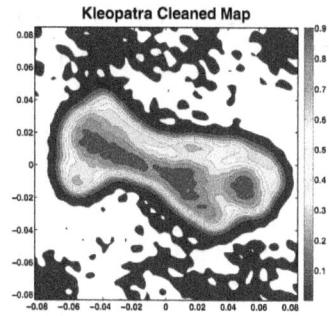

Figure 6.1: Simulated 2012 July 1 04:00 UTC ALMA wide-configuration 750 GHz snapshot image of the asteroid 216 Kleopatra, based on the Ostro et al. (2000) shape model. The model of Kleopatra is 217 x 94 x 81 km in principal-axis extent, so that it spans 21 beams at this time. The cleaned map reflects the signal strength expected for a rotation-limited integration time of 5 min.

the Earth, 6 mas corresponds to 8 km. To avoid rotation smear, integration times would need to be around 3 minutes on average, implying 40 K/beam sensitivity. Thus each beam on the asteroid will have an SNR of 3-6, sufficient to construct a shape model if the object is observed at a large number of orientations.

When constructing a shape model from thermal images, the shape must be separated from the surface temperature distribution, for example using a first-pass shape model that considers only the silhouette of the object in each image (e.g. Simonelli et al. 1993). For a diffraction-limited shape model, the thermal properties of the surface as a function of position must also be modeled. At 750 GHz, ALMA will be sensitive to the temperature roughly 3 mm (10 wavelengths) beneath the surface. Using the temperature at this depth as a function of time-of-day for each location, the thermal inertia of the surface and the temperature profile with depth can be determined. Observations at lower frequencies can probe deeper and provide the temperature profile directly, at the expense of decreased resolution (Chamberlin et al. 2007).

To survey all objects, main belt and Trojan, large enough for shape estimation by ALMA imaging would require observing roughly 800 objects for ~10 epochs of ~5 minutes each (3 minutes on source, 2 minutes for calibrators and slewing) or a total of about 650 hours. If even a fraction of the possible targets are observed, the Yarkovsky migration main-belt objects could be accurately estimated for many hundreds of targets. In addition, the 0.005" astrometry from properly calibrated ALMA data is approximately 20x finer than that from the best optical data (~0.1", Pascu et al. 2002). As a word of caution, calibration of ALMA for high frequencies and long baselines will require very accurate positions for calibrator sources and reliable atmospheric delay models (ALMA 2004).

Having such data for targets throughout the main belt and the Trojans would make a number of dynamical studies possible. Long-term modeling of asteroid orbits (e.g. Bottke et al. 2007) can only be improved with better knowledge of where the objects are and how their orbits will change. Shorter-term trajectory predictions at 5 mas resolution are sensitive to the individual masses of almost all objects ALMA can resolve. The change in position for a typical main-belt or Trojan object from a 100-km perturber with an approach distance of 0.1 AU is ~10 mas/year, and such encounters occur every few years for any given object. Even if only objects larger than 100 km are observed (of which there are about 100, requiring roughly 65 hours of telescope time), ALMA astrometry can be used to estimate their masses. Accurate masses for the largest objects in the main belt would both give a better estimate of the total mass of the belt and provide the densities of individual objects when combined with shape models.

ALMA also has some utility when applied to near-Earth objects. Currently the largest source of uncertainty in trajectory prediction of a near-Earth asteroid, after Yarkovsky migration and radiation pressure, is uncertainty in the positions and masses of other asteroids. ALMA observations of the main belt will decrease the last. In one case ALMA can also remove the first: observations of 1950 DA over several years starting in 2012 can potentially measure the Yarkovsky offset in its position, resolving the ambiguity in its spin state that prevents forecasting the outcome of the 2880 encounter. ALMA cannot replace radar ranging and velocity measurements; a near-Earth asteroid close enough to be detected by ALMA will almost always be detectable by radar at finer resolution; but it will be the most effective tool for a few objects and may be a useful supplement to delay-Doppler and radar-interferometric imaging.

Even if we consider only ALMA-derived shape models without astrometry, the shapes of individual main-belt objects will provide interesting dynamical results. The spin states of asteroids are variable due to YORP realignment, in turn feeding into orbital dynamics, since the Yarkovsky perturbation is determined by spin state. With surface temperature and near-surface temperature profile mapped out across the surface as an object rotates, it will be possible to accurately model the YORP torques. I can then consider the evolution of all of the large main-belt objects with time and estimate how often they are disrupted by spin-up.

In addition to mapping three-dimensional shapes and surface temperature, ALMA could be used to map surface composition. Lovell (2007) considered object-integrated lightcurves or relatively large beams as a means of increasing signal-to-noise. A large beam size is counterproductive for continuum imaging, but is essential for spectral observations. An ALMA beam 12.5 mas across, 20 km at 2 AU, would have SNR of ~5 for a 100-MHz wide spectral channel (spectral resolution power ~800) with an 8 minute integration, providing regional compositional information for the ~100 targets larger than 100 km across.

The spectral properties of solid geological materials in the terahertz range are very poorly known. Based on broadband photometry, Redman et al. 1998 reported a difference in spectral slope between metallic and non-metallic objects over 150-870 GHz, which they attributed to the different emissivities of metal and rock. Some spectral work on terrestrial soils to detect landmines may be relevant (Du Bosq et al. 2005). If the spectra are favorable, submillimeter spectroscopy could establish the composition of the W-class objects, which are spectrally neutral in the optical and near-infrared (Rivkin et al. 2000). There are ~25 known W-class objects larger than 40 km across. For objects such as 129 Antigone, ALMA spectroscopy might establish why one side of the object is radar-dark while the other is highly radar-reflective (Shepard et al. 2008).

ALMA is not yet capable of imaging anything other than constant point sources – there are only a small number of antennas installed and the array will not be available for full science operations until 2012. As a precursor, I have started a project to image Vesta, Juno, and Hygiea with the smaller CARMA array (Woody et al. 2004). In its wide configuration, CARMA has resolution ~0.1" – somewhat coarser than adaptive optics, but marginally sufficient to resolve these three asteroids. CARMA has observed all of them, and my analysis of the data is ongoing.

To provide reassurance that speckle does not trouble observations with ALMA or CARMA, recall that in radar astronomy the phases of all of the photons re-radiated a given point on the surface is constant. Because the thermal-emission photons from each point have random phase, there is no speckle pattern and interferometry and self-calibration will be successful. While interferometry is not quite yet ready to be applied to asteroids, it will become a very powerful tool in the next several years.

6.d. A final thought.

These potential future applications of delay-Doppler imaging, radar speckle tracking, and interferometry all illustrate a shift in our understanding of asteroids. As many more objects are resolved and studied, we will be able to understand the physical properties of the asteroid population, rather than just a handful of objects.

References

ALMA Collaboration, 2004, *ALMA Scientific Specifications and Requirements,* available at http://www.cv.nrao.edu/naasc/.

ALMA Collaboration, 2006, *ALMA Sensitivity*, available at http://www.cv.nrao.edu/naasc/alma_sensitivity.shtml.

Asphaug, E., Nolan, M.C., 1992, *Analytical and numerical predictions for regolith production on asteroids*, LPSC **23**, 43.

Asphaug, E., 2008, *Critical crater diameter and asteroid impact seismology*, Meteori. & Planet. Sci. **43**, 1075-1084.

Baker, J., Bizzarro, M., Wittig, N., Connelly, J., Haack, H., 2005, *Early planetesimal melting from an age of 4.5662 Gyr for differentiated meteorites*, Nature **436**, 1127-1131.

Bardwell, C.M., 2001, Minor Planet Electron. Circ. 2001-A26.

Barucci, M.A., Cruikshank, D.P., Mottola, S., Lazzarin, M., 2002, *Physical properties of Trojan and Centaur asteroids*, in Asteroids III (Bottke, W.F., Cellino, A., Paolicchi, P., Binzel, P.R., eds.), Univ. of Arizona Press, Tucson, 273-287.

Benner, L.A.M., Ostro, S.J., Giorgini, J.D., Busch, M.W., Rose, R., Jao, J.S., Jurgens, R.F., 2007a, *Radar observations of asteroid 2004 XP14: an outlier in the near-Earth population*, Bulletin of the American Astronomical Society **38**, 621.

Benner, L.A.M., and 11 colleagues, 2007b, *Radar images of binary near-Earth asteroid 2006 VV2*, Bulletin of the American Astronomical Society **39**, 432.

Benner, L.A.M., and 10 colleagues, 2008, *Near-Earth asteroid surface roughness depends on compositional class*, Icarus **198**, 294-304.

Benner, L.A.M., and 13 colleagues, 2009, *Arecibo and Goldstone radar images of near-Earth asteroid (136849) 1998 CS1*, Bull. Am. Ast. Soc. **41**, 1083.

Benner, L.A.M., 2010, *Asteroid 3-D shape models estimated from delay-Doppler radar data*, http://echo.jpl.nasa.gov/~lance/shapes/asteroid_shapes.html.

Benson, J.M., 1995, *The VLBA Correlator*, in Very Long Baseline Interferometry and the VLBA, ASP Conf. Series **82**.

Binzel, R.P., Rivkin, A.S., Stuart, J.S., Harris, A.W., Bus, S.J., Burbine, T.H., 2004, *Observed spectral properties of near-Earth objects*, Icarus **170**, 259-294.

Binzel, R.P., Morbideli, A., Merouane, S., DeMeo, F.E., Birlan, M., Vernazza, P., Thomas, C.A., Rivkin, A.S., Bus, S.J., Tokunaga, A.T., 2010, *Earth encounters as the origin of fresh surfaces on near-Earth asteroids*, Nature **463**, 331-334.

Black, G., Campbell, D.B., Treacy, R., Nolan, M.C., 2005, *Radar-interferometric asteroid imaging using a flexible software correlator,* Bull. Amer. Ast. Soc. **37**, 1155.

Booth, R.S., 2002, *The European VLBA network*, Adv. Spac. Res. **11**, 397-401.

Bottke, W.F., Durda, D.D., Nesvorny, D., Jedicke, R., Morbidelli, A., Vokrouhlicky, D., Levison, H., 2005, *The fossilized size distribution of the main asteroid belt*, Icarus **175**, 111-140.

Bottke, W.F., Vokrouhlicky, D., Rubincam, D.P., Nesvorny, D., 2006, *The Yarkovsky and YORP effects: implications for asteroid dynamics*, Ann. Rev. Earth & Planet. Sci. **34**, 157-191.

Bottke, W.F., Vokrouhlicky, D., Nesvorny, D., 2007, *An asteroid breakup 160 Myr ago as the probable source of the K/T impactor*, Nature **449**, 48-53.

Bramson, A., Magri, C., Howell, E.S., Nolan, M.C., Taylor, P.A., 2009, *The Hayabusa spacecraft model of Itokawa: lessons learned for radar shape models*, Bull. Am. Ast. Soc. **41**.

Brigham, E.O., 1974, *The Fast Fourier Transform*, Prentice-Hall, Inc., New Jersey, USA.

Britt, D.T., Consolmagno, G.J., 2003, *Stony meteorite porosities and densities: A review of the data through 2001,* Meteorit. Planet. Sci. **38**, 1161–1180.

Burbine, T.H., McCoy, T.J., Meibom, A., Gladman, B., Keil, K., 2002, *Meteoritic parent bodies: their number and identification,* in Asteroids III (Bottke, W.F., Cellino, A., Paolicchi, P., Binzel, P.R., eds.), Univ. of Arizona Press, Tucson, 653-667.

Bus, S.J., Vilas, F., Barucci, M.A., 2002, *Visible-wavelength spectroscopy of asteroids,* in Asteroids III (Bottke, W.F., Cellino, A., Paolicchi, P., Binzel, P.R., eds.), Univ. of Arizona Press, Tucson, 169-182.

Busch, M.W., 2009, *ALMA and asteroid science,* Icarus **200**, 347-349.

Busch, M.W., Ostro, S.J., Benner, L.A.M., Giorgini, J.D., Jurgens, R.F., Rose, R., Magri, C., Pravec, P., Scheeres, D.J., Broschart, S.B., 2006, *Radar and optical observations and physical modeling of near-Earth asteroid 10115 (1992 SK),* Icarus **181**, 145-155.

Busch, M.W., and 10 colleagues, 2007a, *Arecibo radar observations of Phobos and Deimos,* Icarus **186**, 581-584.

Busch, M.W., and 15 colleagues, 2007b, *Physical modeling of near-Earth asteroid (29075) 1950 DA,* Icarus **190**, 608-612.

Busch, M.W., Kulkarni, S.R., Conrad, A.P., Cameron, P.B., 2007c, *Keck adaptive optics imaging of near-Earth asteroid 2004XP14,* Icarus **189**, 589-590.

Busch, M.W., and 10 colleagues, 2008, *Physical properties of near-Earth asteroid (33342) 1998 WT24,* Icarus **195**, 614-621.

Busch, M.W, Kulkarni, S.R., Conrad, A.P., 2009, *No satellites around 21 Lutetia,* Icarus **203**, 681-682.

Busch, M.W., Kulkarni, S.R., Brisken, W., Ostro, S.J., Benner, L.A.M., Giorgini, J.D., Nolan, M.C., 2010a, *Determining asteroid spin states using radar speckles,* Icarus **in press.**

Busch, M.W., and 14 colleagues, 2010b, *The shape of 2008 EV5: craters and ridges on near-Earth asteroids,* **in prep.**

Campbell, B.A., 2002, *Radar remote sensing of planetary surfaces,* Cambridge Univ. Press, p. 198.

Carilli, C.L., Holdaway, M.A., 1999, *Tropospheric phase calibration in millimeter interferometry,* Radio Sci. **34**, 817-840.

Carrier, W.D., Mitchell, J.K., Mahmood, A., 1973, *The relative density of lunar soil,* Proc. Lunar Sci. Conf. **4**, 2403.

Chamberlin, M.A., Lovell, A.J., Sykes, M.V., 2007, *Submillimeter lightcurves of Vesta,* Icarus **192**, 448–459.

Chesley, S.R., and 9 colleagues, 2003, *Direct detection of the Yarkovsky effect by radar ranging to asteroid 6489 Golevka,* Science **302**, 5651, 1739-1742.

Clark, B.G., 1980, *An efficient implementation of the algorithm "CLEAN",* Astron. Astrophys. **89**, 377-378.

Cohen, M.N., 1991. *An overview of high range resolution radar techniques,* Proc. Nat. Telesys. Conf. 1991, **1**, 107-115.

Cornwell, T., Braun, R., Briggs, D.S., 1999, *Deconvolution,* in Taylor et al. 1999.

Cuk, M., 2007, *Formation and destruction of small binary asteroids,* ApJ **659**, L57-L60.

Cuk, M., Nesvorny, D., 2010, *Orbital evolution of small binary asteroids,* Icarus **in press.**

Davis, D.R., Housen, K.R., Greenberg, R., 1981, *The unusual dynamical environment of Phobos and Deimos,* Icarus **47**, 220–233.

Day, J.M.D., Ash, R.D., Liu, Y., Bellucci, J.J., Rumble, D., McDonough, W.F., Walker, R.J., Taylor, L.A., 2009, *Early formation of evolved asteroidal crust,* Nature **457**, 179-183.

De Pater, I., Kurth, W.S., 2007, *The Solar System at radio wavelengths,* Encyclopedia of the Solar System, 695-718.

Deller, A.T., Tingay, S.J., Bailes, M., West, C, 2007, *DiFX: A software correlator for very long baseline interferometry using multiprocessor computing environments*, PASP **119**, 318-336.

Demura, H., and 19 colleagues, 2006, *Pole and global shape of 25143 Itokawa*, Science **313**, 1347-1349.

Di Martino, M., and 17 colleagues, 2004, *Results of the first Italian planetary radar experiment*, Planet. Space Sci. **52**, 325-330.

Drummond, J.D., Cocke, W.J., Hege, E.K., Strittmatter, P.A., Lamber, J.V., 1985, *Speckle interferometry of asteroids. I. 433 Eros*, Icarus **61**, 132-151.

Du Bosq, T.W., and 9 colleagues, 2005, *Terahertz and millimeter wave transmission of soils*, Terahertz and Gigahertz Electronics and Photonics IV, Proc. SPIE vol. 5727.

Durech, J., Scheirich, P., Kaasalainen, M., Grav, T., Jedicke, R., Denneau, L., 2007, *Physical models of asteroids from sparse photometric data*, IAU Proc. **236**, 191-200.

Fujiwawa, A., and 21 colleagues, 2006, *The rubble-pile asteroid Itokawa as observed by Hayabusa*, Science **312**, 1330-1334.

Galad, A., Vilagi, J., Kornos, L., Gajdos, S., 2009, *Relative photometry of nine asteroids from Modra*, Minor Planet Bull. **36**, 116-118.

Garvin, J.B., Head, J.W., Pettengill, G.H., Zisk, S.H., 1985, *Venus global radar reflectivity and correlations with elevation*, J. Geophys. Res. B **90** (8), 6859–6871.

Giacconi, R., Lo, K.Y., Napier, P., Perley, R.A., Ulvestad, J., 2004. *The Very Large Array Expansion Project Phase II: Completing the EVLA*, NSF Proposal #0435595.

Giorgini, J.D., and 13 colleagues, 2002, *Asteroid 1950 DA's encounter with Earth in 2880: physical limits of collision probability prediction*, Science **296**, 5565, 132-136.

Giorgini, J.D., Benner, L.A.M., Ostro, S.J., Nolan, M.C., Busch, M.W., 2008, *Predicting the Earth encounters of (99942) Apophis*, Icarus **193**, 1-19.

Giorgini, J.D., and 20 colleagues, 2009, *Radar astrometry of small bodies: detection, characterization, trajectory prediction, and hazard assessment*, white paper submitted to the Planetary Science 2010 Decadal Survey.

Green, P.E., 1968, *Radar measurements of target scattering properties*, In Radar Astronomy (Evans, J.V., Hagfors, T., eds.), McGraw-Hill, New York, 1-77.

Gold, T., 1962, *Processes on the lunar surface*, In: Kopal, Z., Mikhailov, Z.K. (Eds.), The Moon. In: *IAU Symposium*, vol. 14. Lunar and Planetary Institute, Houston, TX, pp. 433–440.

Goldreich, P., Sari, R., 2009, *Tidal evolution of rubble piles*, ApJ **691**, 54-60.

Greisen, E.W., 2002, *AIPS, the VLA, and the VLBA*, in *Information Handling in Astronomy*, Astrophys. Space Sci. Library **285**, 109-125.

Gurwell, M., 2004, *Solar System Science with the ALMA*, ALMA Science Workshop, May 2004.

Hamilton, D.P., Krivov, A.V., 1996, *Circumplanetary dust dynamics: Effects of solar gravity, radiation pressure, planetary oblateness, and electromagnetism*, Icarus **123**, 503–523.

Harris, A.W., 1996, *The rotation rates of very small asteroids: evidence for 'rubble pile' structure*, Lunar & Planet. Sci. **27**, 493.

Harris, A.W., 2008, *What Spaceguard did*, Nature **453**, 1178-1179.

Harris, A.W., Mueller, M., Delbo, M., Bus, S.J., 2007, *Physical characterization of the potentially hazardous high-albedo asteroid (33342) 1998 WT24 from thermal-infrared observations*, Icarus **188**, 414-424.

Harris, A.W., Warner, B.D., 2010, *Minor planet lightcurve parameters*, http://cfa-www.harvard.edu/iau/lists/LightcurveDat.html.

Helin, E., 1992, 1992 SK, IAU Circular 5628.

Herschel, W., 1802, *Observations on the two lately discovered celestial bodies*, Phil. Trans. Royal Society. **92**, 213-232.

Hirabayashi, H. and 54 colleagues, 2000, *The VLBI Space Observatory Programme and the Radio-Astronomical Satellite HALCA*, Publ. of the Astron. Soc. of Japan **52**, 955-965.

Hjellming, R.M., 2000, *Radio Astronomy*, in Allen's Astrophysical Quantities (Cox, A.N.,ed.), Springer-Verlag Press, New York, 121-142.

Holsapple, K.A., 2004, *Equilibrium figures of spinning bodies with self-gravity*, Icarus **172**, 272–303

Hsieh, H.H., Jewitt, D., 2006, *A population of comets in the main asteroid belt*, Science **312**, 561-563.

Hudson, R. S., 1993, *Three-dimensional reconstruction of asteroids from radar observations*, Remote Sensing Reviews **8**, 195-203.

Hudson, R. S., Ostro, S. J., 1995, *Shape and non-principal axis spin state of asteroid 4179 Toutatis*. Science **270**, 84-86.

Hudson, R. S., Ostro, S.J., Scheeres. D.J., 2003, *High-Resolution Model of Asteroid 4179 Toutatis*. Icarus **161**, 346-355.

Ivezic, Z, Tyson, J.A., Juric, M., Kubica, J., Connolly, A., Pierfederici, F., Harris, A.W., Bowell, E., LSST Collaboration, 2007, *LSST: Comprehensive NEO detection, characterization, and orbits*, IAU Proceedings **2**, 353-362.

Jedicke, R., Magnier, E.A., Kaiser, N., Chambers, K.C., 2007, *The next decade of solar system discovery with Pan-STARRS*, Proc. Internat. Ast. Union 2006 **2**, 341-352.

Jenniskens, P., 1994, *Meteor stream activity I. the annual streams*, Astron. Astrophys. **287**, 990-1013.

Jenniskens, P., and 34 colleagues, 2009, *The impact and recovery of asteroid 2008 TC3*, Nature **458**, 485-488.

Kaasalainen, M., Mottola, S., Fulchignoni, M., 2002, *Asteroid models from disk-integrated data*, in Asteroids III (Bottke, W.F., Cellino, A., Paolicchi, P., Binzel, P.R., eds.), Univ. of Arizona Press, Tucson, 139-150.

Kiselev, N.N., Rosenbush, V.K., Jockers, K., Velichko, F.P., Shakhovskoj, N.M., Efimov, Y.S., Lupishko, D.F., Rumyantsev, V.V., 2002, *Polarimetry of near-Earth asteroid 33342 (1998 WT24)*, In: Warmbein, B. (Ed.), Proc. of the Conference: Asteroids, Comets, Meteors ACM 2002. ESA SP-500. ESA, Noordwijk, The Netherlands, pp. 887–890.

Kholin, I.V., 1988, *Spatial-temporal coherence of a signal diffusely scattered by an arbitrarily moving surface for the case of monochromatic illumination*, Radiophys. & Quant. Elec. **31**, 371-374 (translation from Russian original).

Kholin, I.V., 1992, *Accuracy of body-rotation-parameter measurement with monochromatic illumination and two-element reception*, Radiophys. & Quant. Elec. **35**, 284-287 (translation from Russian original).

Kholin, I.V., 2004, *Long-range coherence of the radar field scattered by a rotating Mercury*, Solar Sys. Res. **38**, 449-454 (translation from Russian original).

Kirkpatrick, S., Gelatt, C.D., Vecchi, M.P., 1983, *Optimization by simulated annealing*. Science **220**, 671-680.

Kirkwood, D., 1866, *On the Theory of Meteors*, Proc. Amer. Ass. Adv. Sci., 1866, 8-14.

Krivov, A.V., Hamilton, D.P., 1997, *Martian dust belts: waiting for discovery*, Icarus **128**, 335-353.

Krugly, N.N., Y.N., Belskaya, I.N., Chiorny, V.G., Shevchenko, V.G., Gaftonyuk, N.M., 2002, *CCD photometry of near-Earth asteroids in 2001*, in Warmbein, B. (Ed.), Proc. Of ACM 2002, ESA SP-500, Noordwijk, The Netherlands, 903-906.

Kryszczynska, A., La Spina, A., Paolicchi, P., Harris, A.W., Breiter, S., Pravec, P., 2007, New findings on asteroid spin-vector distributions, Icarus **192**, 223-237.

La Spina, A., Paolicchi, P., Kryszczynska, A., Pravec, P., 2004, *Retrograde spins of near-Earth asteroids from the Yarkovsky effect*, Nature **428**, 400-401.

Latham, J.P., Munjiza, A., Lu, Y., 2002, *On the prediction of void porosity and packing of rock particulates,* Powder Technol. **125**, 10–27.

Lazzarin, M., Marchi, S., Barucci, M.A., Di Martino, M., Barbieri, C., 2004, *Visible and near-infrared spectroscopic investigation of near-Earth objects at ESO: first results,* Icarus **169**, 373-384.

Levison, H.F., Bottke, W.F., Gounelle, M., Morbidelli, A., Nesvorny, D., Tsiganis, K., 2009, *Contamination of the asteroid belt by primordial trans-Neptunian objects,* Nature **460**, 364-366.

LONEOS Sky Survey, 2001, Minor Planet Electron. Circ. 2001-A22.

Lovell, A., 2007, *Observations of asteroids with ALMA,* Astrophys. Space Sci., **313,** 191-196.

LSST Science Collaborations, 2010, *LSST Science Book,* available at http://www.lsst.org.

MacNeal, B.E., Abraham, D.S., Cesarone, R.J., 2007, *DNS antenna array architectures based on future NASA mission needs,* IEEEAC paper 1386.

Magri, C., Consolmagno, G.J., Ostro, S.J., Benner, L.A.M., Beeney, B.R., 2001, *Radar constraints on asteroid regolith properties using 433 Eros as ground truth.* Meteorit. Planet. Sci. **36**, 1697–1709.

Magri, C., Ostro, S.J., Scheeres, D.J., Nolan, M.C., Giorgini, J.D., Benner, L.A.M., Margot, J.-L., 2007, *Radar observations and a physical model of asteroid 1580 Betulia,* Icarus **186**, 152-177.

Magri, C., 2010, *SHAPE software introduction and documentation.* Available from C. Magri on request.

Marchis, F., and 17 colleagues, 2006a, *A low density of 0.8 g cm^{-3} for the Trojan binary asteroid 617 Patroclus,* Nature **439**, 565-567.

Marchis, F., Kaasalainen, M., Hom, E.F.Y., Berthier, J., Enriquez, J., Hestroffer, D., Le Mignant, D., de Pater, I., 2006b, *Shape, size and multiplicity of main-belt asteroids: I. Keck adaptive optics survey,* Icarus **185**, 39-63.

Marchis, F., Descamps, P., Berthier, J., Emery, J.P., 2008, *IAU Circular 8980: S/2008 (216) 1 and S/2008 (216) 2.*

Margot, J.-L., Nolan, M.C., Benner, L.A.M., Ostro, S.J., Jurgens, R.F., Giorgini, J.D., Slade, M.A., Campbell, D.B., 2002, *Binary asteroids in the near-Earth object population,* Science **296**, 1445-1448.

Margot, J.-L., Brown, M.E., 2003, *A low-density M-type asteroid in the main belt,* Science **300**, 1939-1942.

Margot, J.-L., Peale, S.J., Jurgens, R.F., Slade, M.A., Holin, I.V., 2007, *Large longitude libration of Mercury reveals a molten core,* Science **316**, 710-714.

Muhleman, D.O., Butler, B.J., Grossman, A.W., Slade, M.A., 1991, *Radar images of Mars,* Science **253**, 1508-1513.

Napier, P.J., Thompson, A.R., Ekers, R.D., 1983, *The Very Large Array: design and performance of a modern synthesis radio telescope,* IEEE Proceedings **71**, 1295-1320.

Napier, P.J., Bagri, D.S., Clark, B.G., Rogers, A.E.E., Romney, J.D., Thompson, A.R., Walker, R.C., 1994, *The Very Long Baseline Array,* IEEE Proceedings **82**, 5, 658-672.

NEODys, 2010, *Near-Earth objects – dynamics,* http://newton.dm.unipi.it/neodys/

Nesvorny, D., Enke, B.L., Bottke, W.F., Durda, D.D., Asphaug, E., Richardson, D.C., 2006, *Karin cluster formation by asteroid impact,* Icarus **183**, 296-311.

Noll, K.S, Weaver, H.A., Storrs, A.D., Zellner, B., 1995, *Imaging of asteroid 4179 Toutatis with the Hubble Space Telescope,* Icarus, **113**, 353-359.

O'Brien, D.P., Greenberg, R., 2005, *The collisional and dynamical evolution of the main-belt and NEA size distributions,* Icarus, **178**, 179-212.

Office of Spectrum Management, US Department of Commerce, 2003, *United States Frequency Allocations.*

Öpik, E.J., 1951, *Collision probabilities with the planets and the distribution of interplanetary matter,* Proceedings of the Royal Irish Academy **54**A, 165–199.

Ostro, S.J., Campbell, D.B., Shapiro, I.I., 1985, *Main belt asteroids: Dual polarization radar observations*, Science **229**, 442–446.

Ostro, S.J., Jurgens, R.F., Yeomans, D.K., Standish, E.M., Greiner, W., 1989, *Radar detection of Phobos*, Science **243**, 1584–1586.

Ostro, S.J., 1993, *Planetary radar astronomy*, Reviews of Modern Physics **65**, 4, 1235 – 1279.

Ostro, S. J., 1997, *Radar reconnaissance of near-Earth objects at the dawn of the next millennium*, Annals New York Academ. Sci. **822**, 118-139.

Ostro, S.J., 2007, *Planetary radar*, in *Encyclopedia of the solar system* (McFadden, Weissman, Johnson, eds.), Academic Press, 735-764.

Ostro, S.J., and 19 colleagues, 1999, *Radar and optical observations of asteroid 1998 KY26*, Science **285**, 557-559.

Ostro, S.J., Hudson, R.S., Nolan, M.C., Margot, J.-L., Scheeres, D.J., Campbell, D.B., Magri, C., Giorgini, J.D., Yeomans, D.K., 2000. *Radar observations of asteroid 216 Kleopatra*, Science **288**, 836-839.

Ostro, S.J., Hudson, R.S., Benner, L.A.M., Nolan, M.C., Giorgini, J.D., Scheeres, D.J., Jurgens, R.F., Rose, R., 2001, *Radar observations of asteroid 1998 ML14*, Meteorit. & Planet. Sci. **36**, 1225-1236.

Ostro, S.J., Hudson, R.S., Benner, L.A.M., Giorgini, J.D., Magri, C., Margot, J.-L., Nolan, M.C., 2002, *Asteroid radar astronomy,* in Asteroids III (Bottke, W.F., Cellino, A., Paolicchi, P., Binzel, P.R., eds.), Univ. of Arizona Press, Tucson 151-168.

Ostro, S.J., and 15 colleagues, 2006, *Radar imaging of binary near-Earth asteroid (66391) 1999 KW4*, Science **314**, 1276-1280.

Ostro, S.J., Magri, C., Benner, L.A.M., Giorgini, J.D., Nolan, M.C., Hine, A.A., Busch, M.W., Margot, J.-L., 2010, *Radar imaging of asteroid 7 Iris*, Icarus **in press.**

Pascu, D., and 9 colleagues, 2002, *Solar System Astrometry,* in *The Future of Solar System Exploration, 2003-2013: Community Contributions to the NRC Solar System Exploration Decadal Survey*, ASP Conference Series **272**, 361-374.

Pearson, T.J., Readhead, A.C.S., 1984, *Image formation by self-calibration in radio astronomy*, Ann. Rev. of Astron. Astrophys. **22**, 97-130.

Petit, J.-M., Morbidelli, A., Chambers, J., 2001, *The primordial excitation and clearing of the asteroid belt*, Icarus **153**, 338-347.

Perley, R.A., Taylor, G.B., 2003, *The VLA calibrator manual*, NRAO internal document. Available from http://www.vla.nrao.edu/astro/calib/manual.

Pravec, P., Wolf, M., Sarounova, L., 1998, *Lightcurves of 26 near-Earth asteroids*, Icarus **136**, 124–153.

Pravec, P., Harris, A.W., 2000, *Fast and slow rotation of asteroids*, Icarus **148**, 12-20.

Pravec, P., Kusnirak, P., Sarounova, L., Harris, A.W., Binzel, R.P., Rivkin, A.S., 2002, *Large coherent asteroid 2001 OE84*, Proc. Asteroids, Comets, and Meteors 2002, 743-745.

Pravec, P., Harris, A.W., 2007, *Binary asteroid population 1. Angular momentum content*, Icarus 190, 250-259.

Pravec, P., 2010, *Ondrejov asteroid photometry project*, available online at www.asu.cas.cz/~ppravec/neo/htm.

Redman, R.O., Feldman, P.A., Matthews, H.E., 1998. *High-quality photometry of asteroids at millimeter and submillimeter wavelengths*, Astronomical Journal **116**, 1478-1490.

Rivkin, A.S., Howell, E.S., Lebofksy, L.A., Clark, B.E., Britt, D.T., 2000. *The nature of M-class asteroids from 3-μm observations,* Icarus **145**, 351-368.

Rivkin, A.S., Brown, R.H., Trilling, D.E., Bell, J.F., Plassmann, J.H., 2002, *Near-infrared spectrophotometry of Phobos and Deimos,* Icarus **156**, 64–75.

Rivkin, A.S., Binzel, R.P., Bus, S.J., 2005, *Constraining near-Earth object albedos using near-infrared spectroscopy,* Icarus **175**, 175-180.

Rubincam, D.P., 1995, *Asteroid orbit evolution due to thermal lag,* JGR **100**(E1), 1585-1594.

Rubincam, D.P., 2000, *Radiative spin-up and spin-down of small asteroids,* Icarus **148**, 2-11.

Scheeres, D.J., 2007a, *The dynamical evolution of uniformly rotating asteroids subject to YORP,* Icarus **188**, 430-450.

Scheeres, D.J., 2007b, *Rotational fission of contact binary asteroids,* Icarus **189**, 370-385.

Scheeres, D. J., Ostro, S. J., Hudson, R. S., Werner, R. A, 1996, *Orbits close to asteroid 4769 Castalia,* Icarus **121**, 67-87.

Scheeres, D.J., and 15 colleagues, 2006, *Dynamical configuration of binary near-Earth asteroid (66391) 1999 KW4,* Science **314**, 1280-1283.

Scheeres, D.J., Mirrahimi, S., 2008, *Rotational dynamics of a solar system body under solar radiation torques,* Celest. Mech. & Dynam. Ast. **101**, 69-103.

Schwarz, U.J., 1979, *The method 'CLEAN' - use, misuse and variations,* in *Image Formation from Coherence Functions in Astronomy,* C. van Schooneveld, ed., 261-275.

Shambayati, S., Border, J.S., Morabito, D.D., Mendoza, R., 2007, *MRO Ka-band demonstration: cruise phase lessons learned,* IEEE Aerospace Conference 2007, 1-17.

Shepard, M.K., and 19 colleagues, 2008, *A radar survey of M- and X-class asteroids,* Icarus **195**, 184-205.

Simonelli, D.P., Thomas, P.C., Carcich, B.T., Veverka, J., 1993, *The generation and use of numerical shape models for irregular solar system objects,* Icarus **103**, 49-61.

Skilling, J., Bryan, R.K., 1984, *Maximum entropy image reconstruction: general algorithm,* Mon. Not. R. Astr. Soc. **211**, 111-124.

Slade, M.A., Butler, B.J., Muhleman, D.O., 1992, *Mercury radar imaging: evidence for polar ice,* Science **258**, 635-640.

Slade, M.A., Giorgini, J.D., Preston, R.A., Brozovic, M., 2009, *New developments in DSN radar observations of NEOs,* White Paper submitted to the 2010 Planetary Science Decadal Survey.

Soter, S., 1971, *The dust belts of Mars,* CRSR Report No. 462, Cornell University.

Statler, T.S., 2009, *Extreme sensitivity of the YORP effect to small-scale topography,* Icarus **202**, 502-513.

Taylor, G.B., Carilli, C.L., Perley, R.A., editors. 1999, *Synthesis Imaging in Radio Astronomy II,* Astronomical Society of the Pacific Conference Series, vol. 180.

Taylor, P.A., and 11 colleagues, 2007, *Spin rate of asteroid (54509) 2000 PH5 increasing due to the YORP effect,* Science **316**, 274-277.

Thomas, P., Veverka, J., Dermott, S., 1986, *Small satellites,* In: Burns, J.A., Matthews, M.S. (Eds.), *Satellites,* Univ. of Arizona Press, Tucson, 802–835.

Thomas, N., Britt, D.T., Herkenhoff, K.E., Murchie, S.L., Semenov, B., Keller, H.U., Smith, P.H., 1999, *Observations of Phobos, Deimos, and bright stars with the imager for Mars Pathfinder,* J. Geophys. Res. **104**, 9055–9068.

Thompson, A.R., Clark, B.G, Wade, C.M., Napier, P.J., 1980, *The Very Large Array,* Astrophys. Jour. Supp. **44**, 151-167.

Thompson, A.R., 1999, *Fundamentals of Radio Interferometry,* in *Synthesis Imaging In Radio Astronomy II,* Astronomical Society of the Pacific Conference Series **180**, 11-36.

Veverka, J., Thomas, P., Johnson, T.V., Matson, D., Housen, K., 1986, *The physical characteristics of satellite surfaces,* In: Burns, J.A., Matthews, M.S.(Eds.), *Satellites,* Univ. of Arizona Press, Tucson, 342–402.

Veverka, J., and 13 colleagues, 1999, *NEAR encounter with asteroid 253 Mathilde: overview*, Icarus **140**, 3-16.

Veverka, J., and 32 colleagues, 2000, *NEAR at Eros: imaging and spectral results*, Science **289**, 2088-2097.

Walsh, K.J., Richardson, D.C., Michel, P., 2008, *Rotational breakup as the origin of small binary asteroids*, Nature **454**, 188-191.

Whiteley, R.J., Hergenrother, C.W., Tholen, D.J., 2002, *Monolithic fast-rotating asteroids*, Proc. Asteroids, Comets, Meteors 2002, 473-480.

Wirtanen, C.A., 1950, Minor Planet Circ. 416.

Wizinowich, P., Acton, D.S., Shelton, C., Stomski, P., Gathright, J., Ho, K., Lupton, W., Tsubota, K., 2000, *First light adaptive optics images from the Keck II telescope*, Publ. Astron. Soc. Pac. **112**, 315-319.

Woody, D.P., and 11 colleagues, 2004, *CARMA: a new heterogeneous millimeter-wave interferometer*, Proc. SPIE **5498**, 30

Yeomans, D.K., and 15 colleagues, 2000, *Radio science results during the NEAR-Shoemaker spacecraft rendezvous with Eros*, Science **289**, 2085-2088.

Zuber, M.T., and 11 colleagues, 2002, *The shape of 433 Eros from the NEAR-Shoemaker Laser Rangefinder*, Science **289**, 2097-2101.

Appendix 1: Delay Doppler Radar Images Used In Shape Modeling

Appendix 1.a. 1998 WT24

1998 WT24 Master Log

Date	Start	Stop	Observatory		Resolution	Runs	Sub-Lat °	RA °	DEC °	RTT s
2001 Dec 14	06:35:54	07:28:55	Goldstone		0.25usx1.0Hz	23				
2001 Dec 14	08:17:10	08:31:00	Goldstone	Rb	0.125usx0.5Hz	25	−5	99	29	15
2001 Dec 14	12:46:50	14:05:20	Goldstone	Rb	0.125usx0.5Hz	27				
2001 Dec 15	04:43:29	13:05:10	Goldstone	Lb	0.125usx0.5Hz	39	−15	82	36	13
2001 Dec 16	01:37:42	05:00:02	Goldstone	Lb	0.125usx0.5Hz	41	−24	60	41	12
2001 Dec 16	09:17:43	11:10:15	Goldstone	L	0.125usx0.5Hz	41	−28	51	42	12
2001 Dec 17	02:03:20	09:00:00	Goldstone	Lb	0.125usx0.25Hz	43	−35	28	41	13
2001 Dec 18	01:32:05	07:30:24	Goldstone	Lb	0.125usx0.25Hz	45				
2001 Dec 18	21:08:49		Goldstone	Lb	0.125usx0.25Hz	49	−39	358	31	19
2001 Dec 19		06:40:31	Goldstone							
2001 Dec 15	04:09:08	05:17:53	Arecibo	L	0.1usx0.06Hz	39	−15	82	36	13
2001 Dec 17	23:03:21		Arecibo	L	0.05usx0.09Hz	39	−38	11	37	15
2001 Dec 18		00:00:42	Arecibo							
2001 Dec 18	22:54:05	23:03:53	Arecibo	L	0.05usx0.07Hz	47	−39	358	31	19
2001 Dec 19	21:22:17	23:04:40	Arecibo	L	0.1usx0.06Hz	47	−38	350	27	19
2001 Dec 20	20:12:52	22:28:44	Arecibo	L	0.1usx0.05Hz	47	−38	345	23	22

Radar observations of WT24 used in shape modeling. Sub-Lat is the latitude of the sub-radar point at the mid-point of the observations. Similarly, the asteroid's sky position and round-trip travel time (RTT) are given at the mid-point of each track. All the Goldstone tracks span at least one complete rotation and the last crosses a day boundary. 'b' denotes a bistatic observation, with transmit at one station and receive at another. 'R' and 'L' denote the right and left senses of circular polarization transmitted.

The following collage shows all radar images used in fitting and corresponding model images and plane-of-sky views. The format for each row is, from left to right: observed image, fit, prograde model. Delay-Doppler projections have range increasing from top and Doppler from left. Time increases downward and from left within each day.

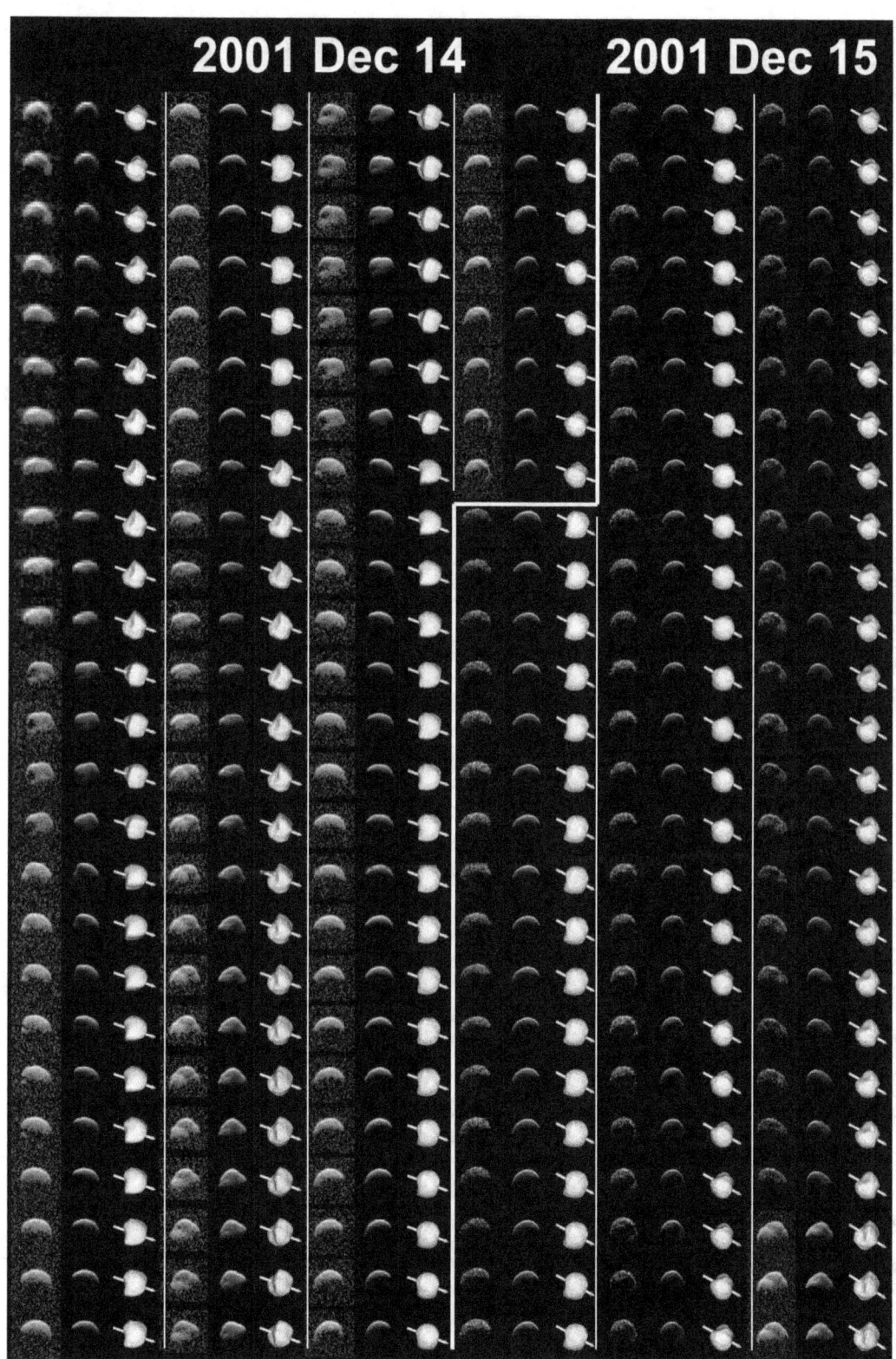

2001 Dec 14   2001 Dec 15

1998 WT24 image collage, page 1 of 6

81

2001 Dec 16

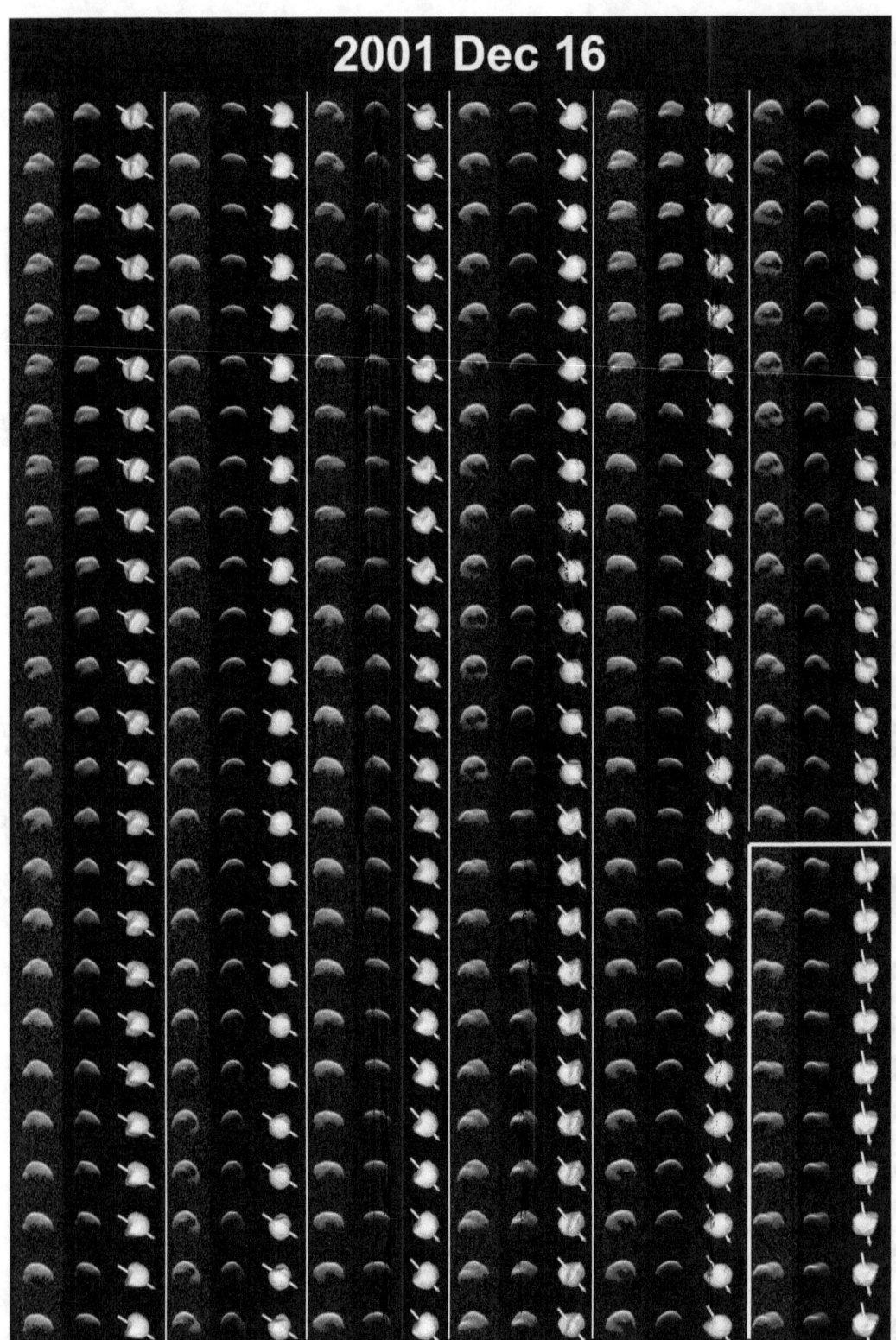

1998 WT24 image collage, page 2 of 6

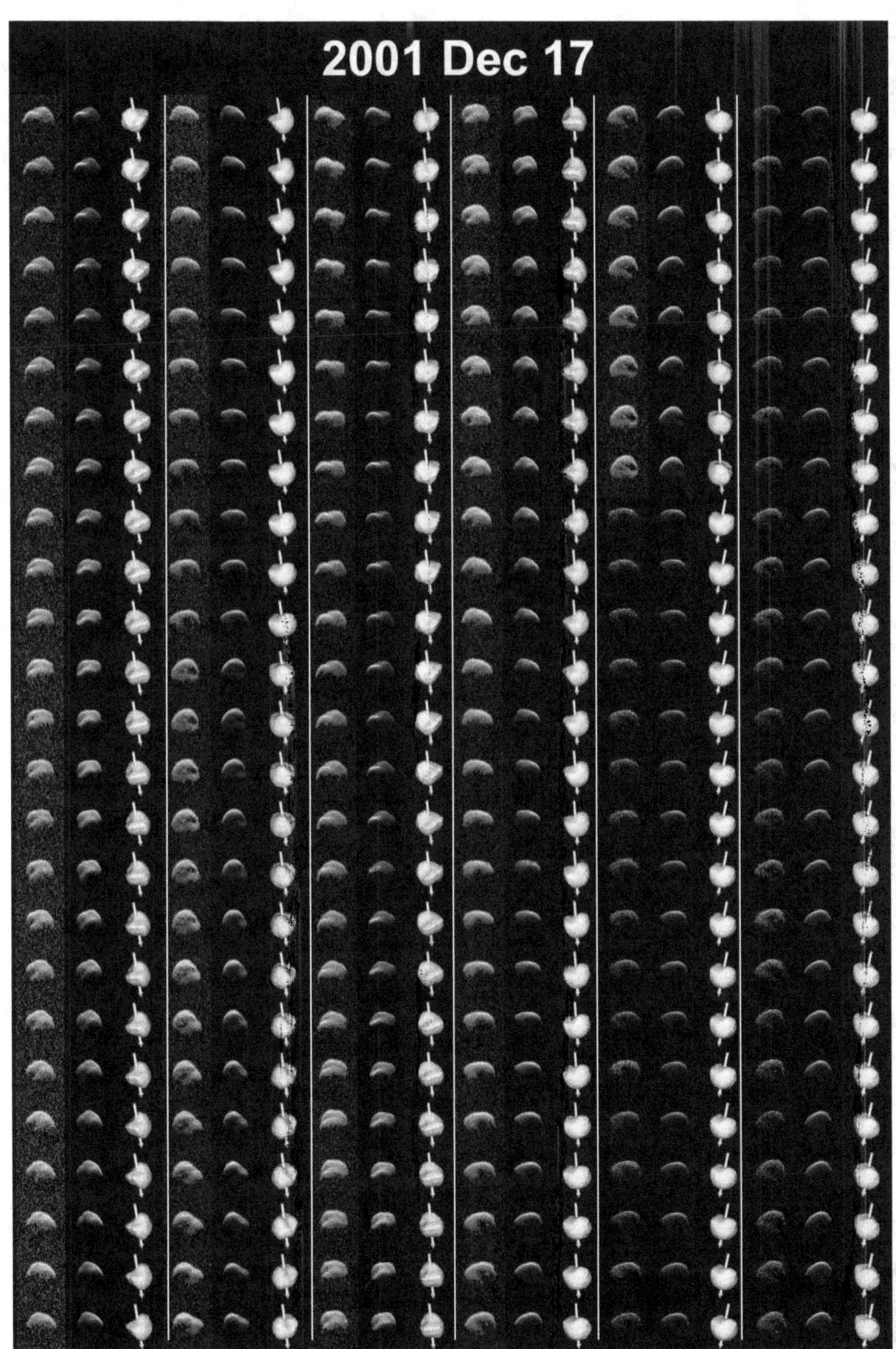

2001 Dec 17

1998 WT24 image collage, page 3 of 6

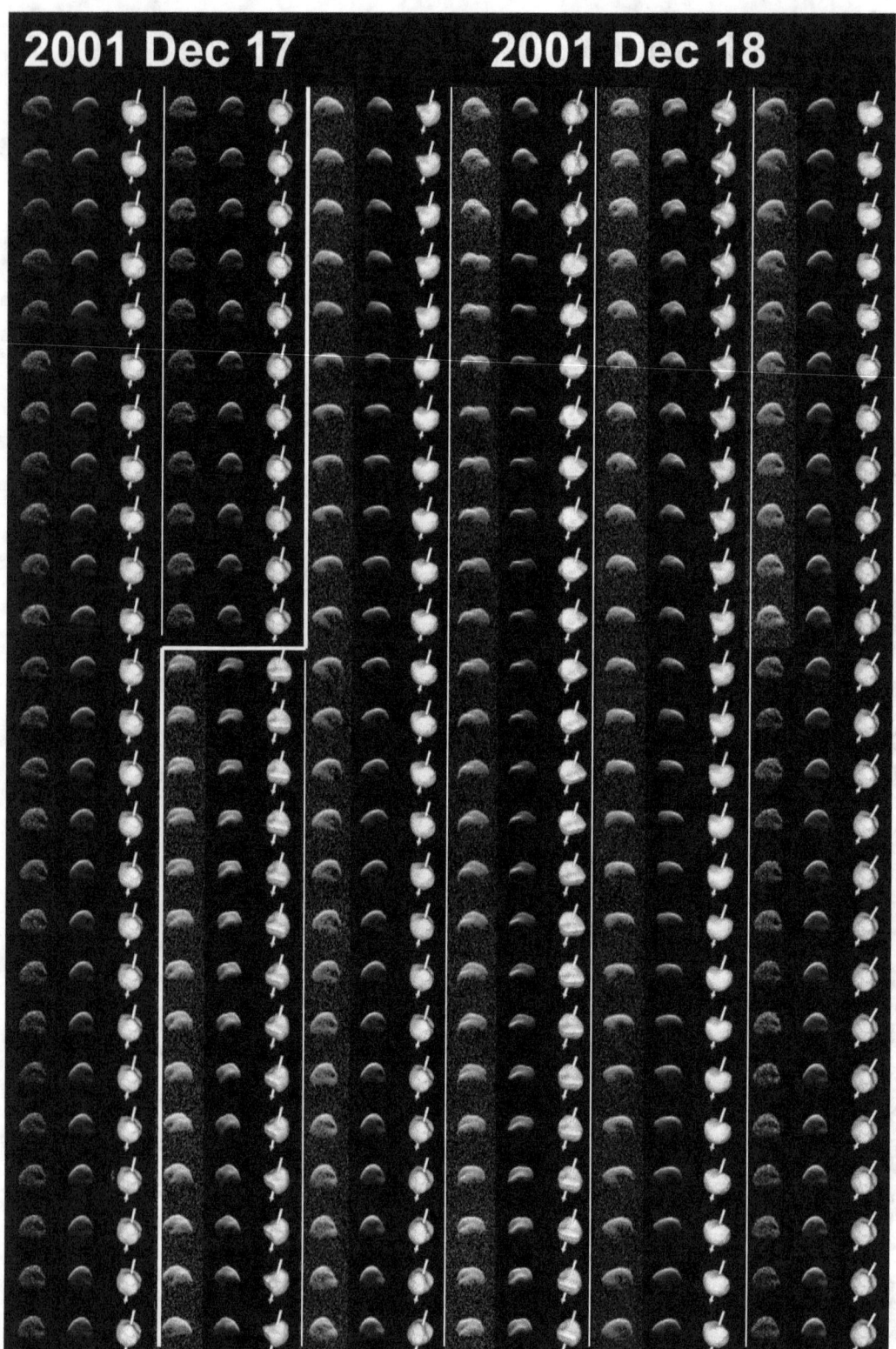

1998 WT24 image collage, page 4 of 6

1998 WT24 image collage, page 5 of 6

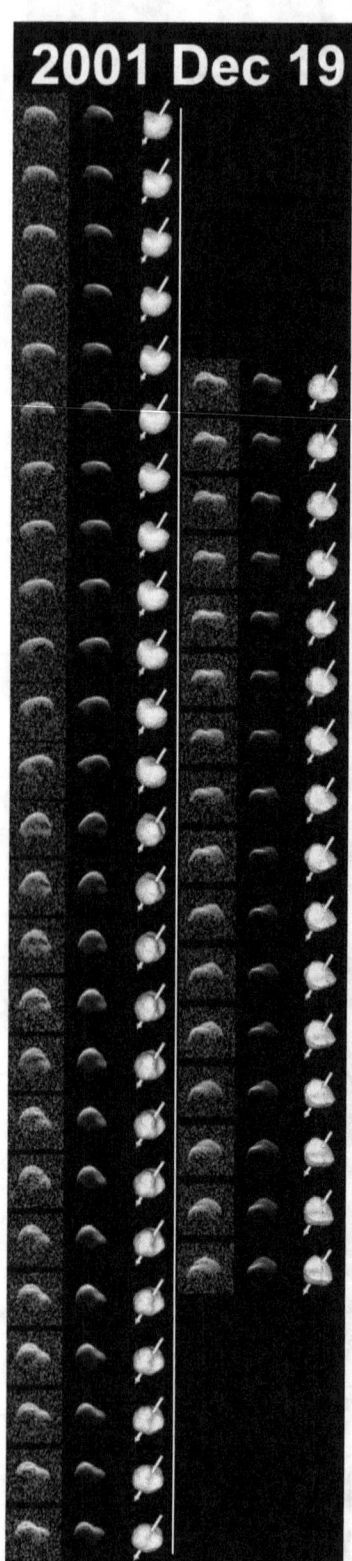

2001 Dec 19

1998 WT24 image collage, page 6 of 6

1950 DA Master Log

Ondrejov Lightcurves:

Time (UTC)	Time (MJD)	Number Points	Rotation Phase	Rotations	Sub-Lat Pro	Ret
2001 Jan 29, 03:53 — 05:29	51938.162 — 51938.229	26	237° — 147°	0.75	−39°	37°
2001 Feb 15, 01:32 — 04:55	51955.064 — 51955.205	83	195° — 49°	1.59	−47°	47°
2001 Feb 15, 20:32 — Feb 16, 04:36	51955.856 — 51956.192	181	179° — 108°	3.80	−47°	48°
2001 Feb 28, 02:41 — 04:50	51968.112 — 51968.202	169	52° — 56°	1.01	−53°	63°

Radar:

Date	Observatory	Resolution	Time (UTC)	Runs	Rotation Phase	Rotations	Sub-Lat Pro	Ret
2001 Mar 3	Arecibo	0.10µs x 0.125Hz	11:27:51 — 11:59:32	18	79° — 168°	0.25	−43°	55°
2001 Mar 3	Goldstone	1.00µs x 6.152Hz	15:32:56 — 15:55:54	13	52° — 117°	0.18	−42°	54°
2001 Mar 3	Goldstone	0.25µs x 3.001Hz	16:04:47 — 18:14:45	61	142° — 149°	1.02	−42°	54°
2001 Mar 4	Goldstone	0.25µs x 3.001Hz	10:09:56 — 11:30:38	36	356° — 199°	0.63	−37°	49°
2001 Mar 4	Arecibo	0.500Hz	10:39:41 — 11:08:17	13	55° — 136°	0.22	−37°	49°
2001 Mar 4	Arecibo	0.10µs x 0.256Hz	11:10:41 — 12:00:04	28	143° — 282°	0.39	−37°	49°
2001 Mar 7	Goldstone	1.00µs x 6.152Hz	12:11:30 — 16:31:17	115	292° — 250°	2.04	−16°	26°

Log of observations used in the 1950 DA shape modeling, giving observatory, observation resolution (frequency-only denotes CW), track duration, rotation phase coverage, sub-Earth latitude at the midpoint of each set of observations for both of our models, and, for the radar observations, the number of transmit-receive cycles (runs). Rotation phase is relative to 2001 March 3, 11:00:00 UT. Goldstone observations spanned more than a complete rotation on March 3 and 7. All transmissions were left-hand circular polarized.

The following collage contains the radar images of DA used in the shape modeling, and corresponding fits and plane-of-sky projections of the prograde and retrograde model, grouped by day and observatory. The format for each row is, from left to right: retrograde model, retrograde fit, observed image at scale of retrograde fit, observed image at scale of prograde fit, prograde fit, prograde model. Delay-Doppler projections have range increasing from top and Doppler from left. Time increases downward and from left within each day.

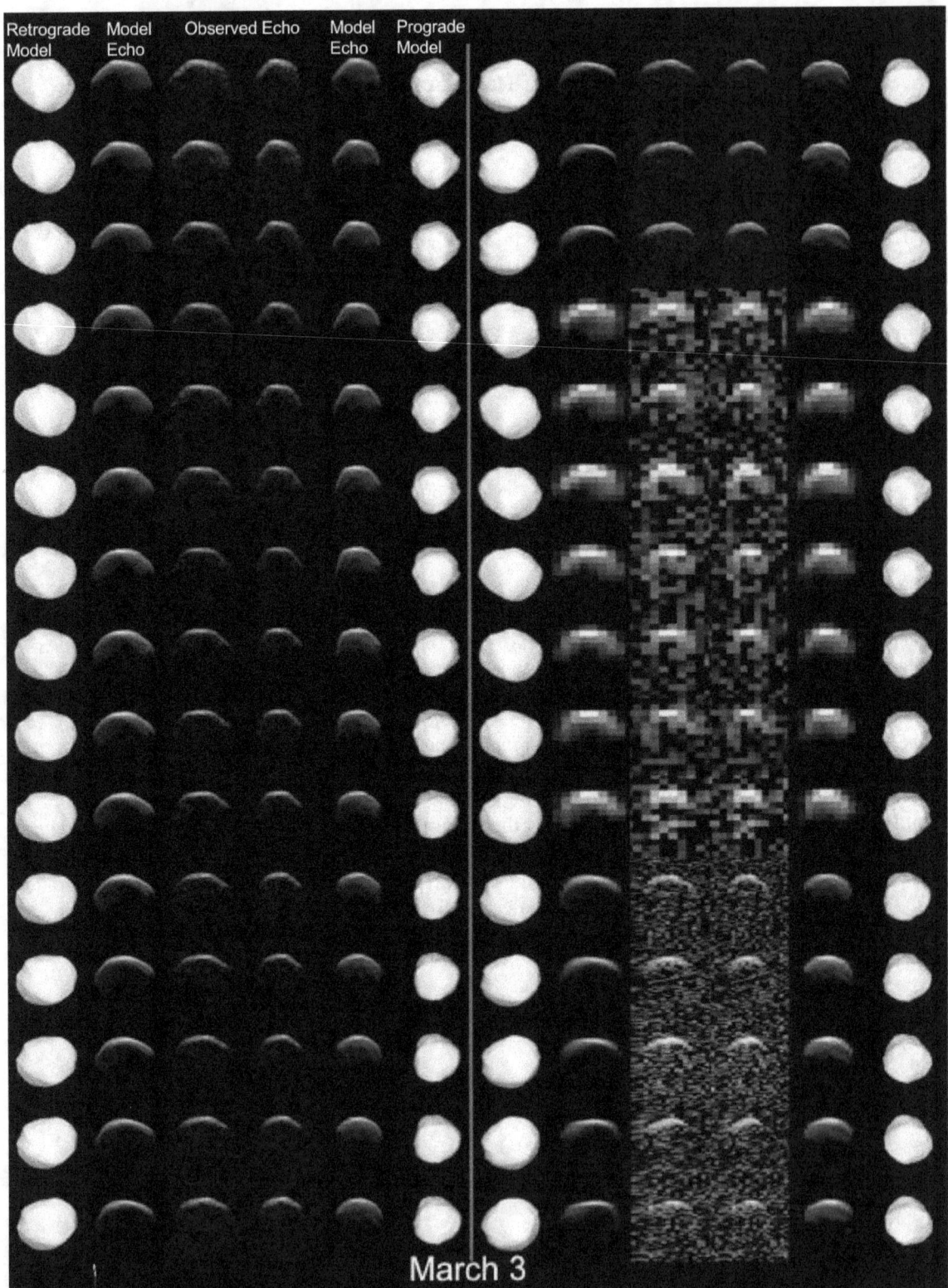

Retrograde Model | Model Echo | Observed Echo | Model Echo | Prograde Model

March 3

1950 DA image collage, page 1 of 4

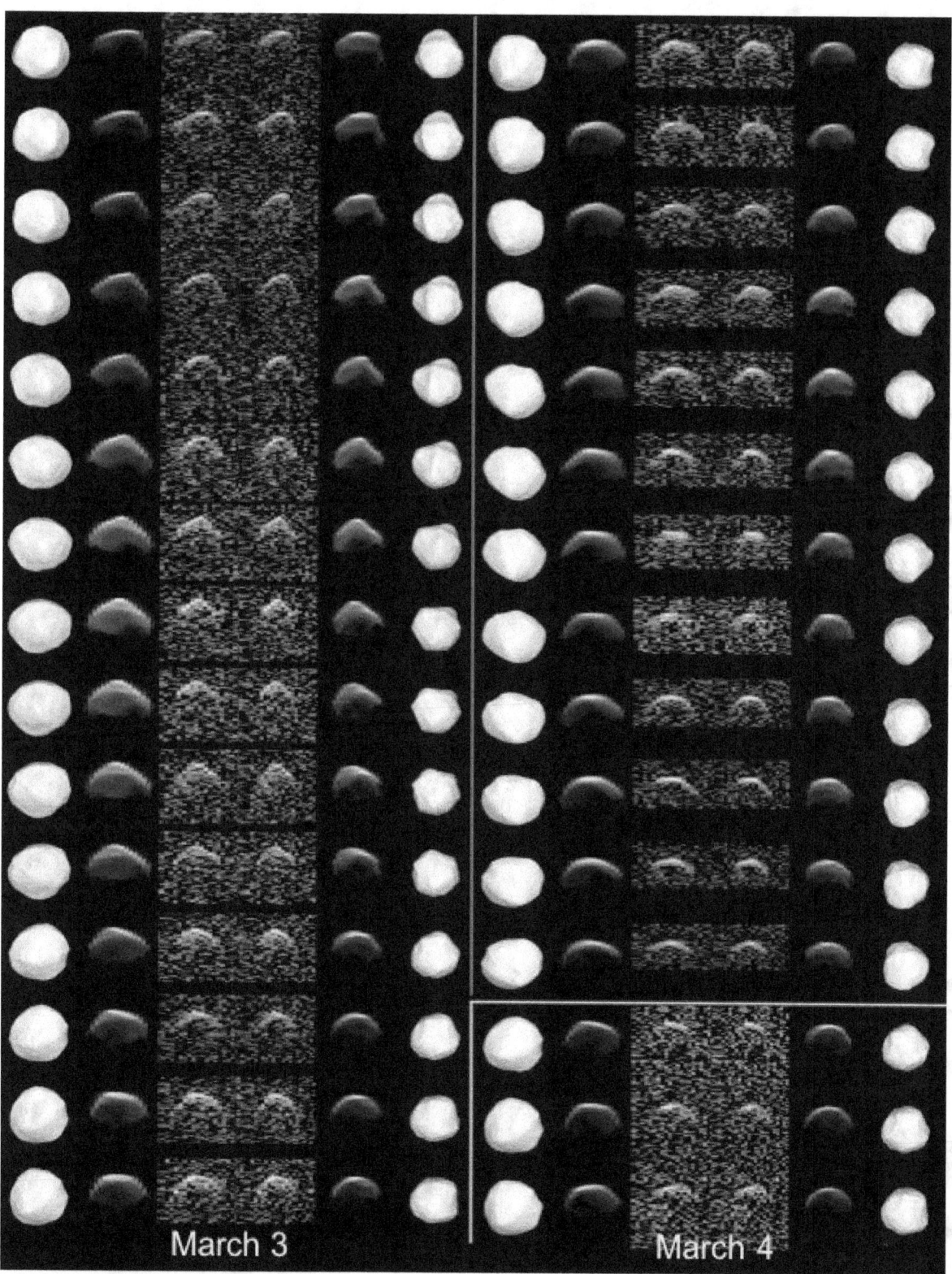

March 3

March 4

1950 DA image collage, page 2 of 4

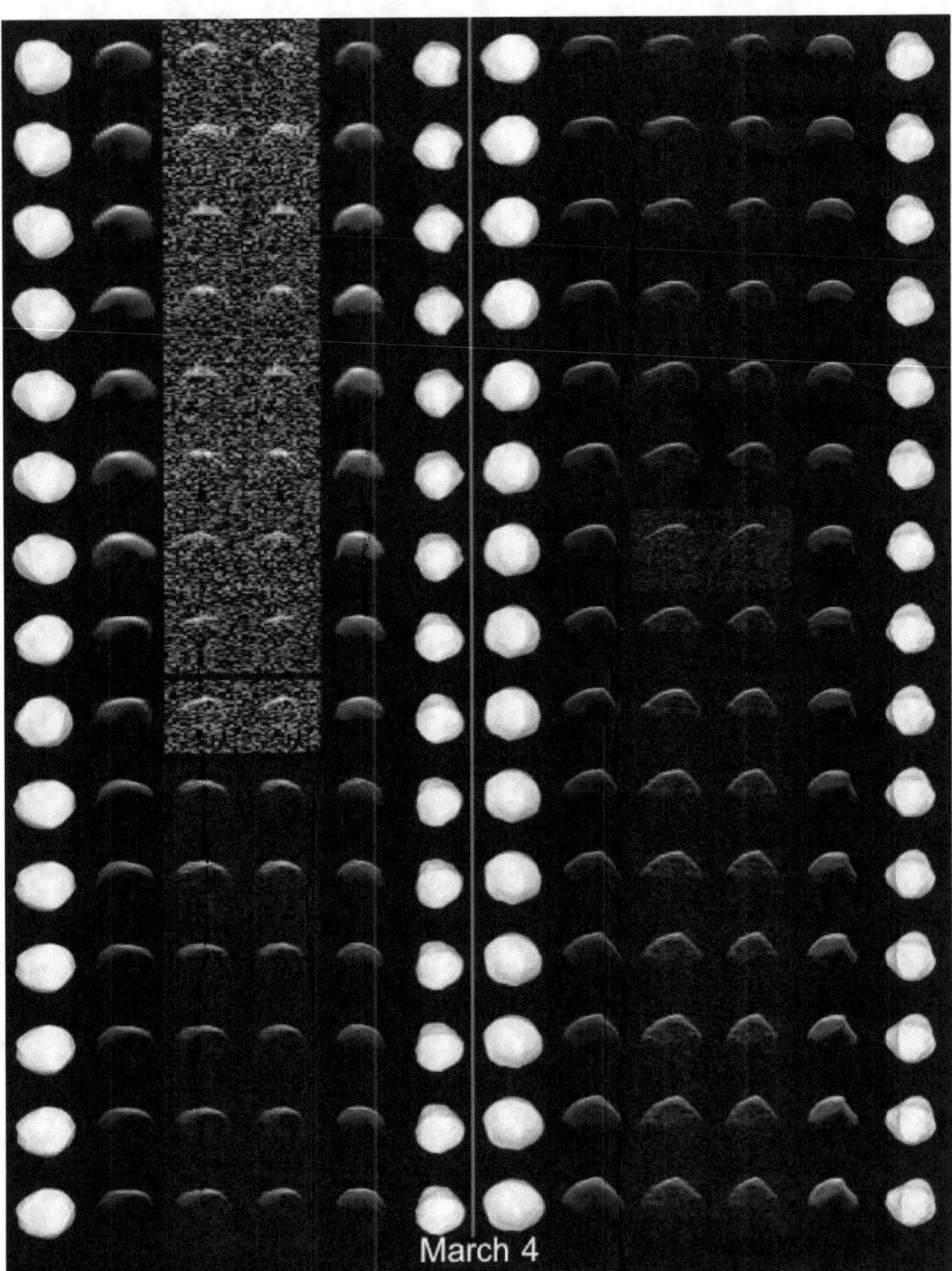

March 4

1950 DA image collage, page 3 of 4

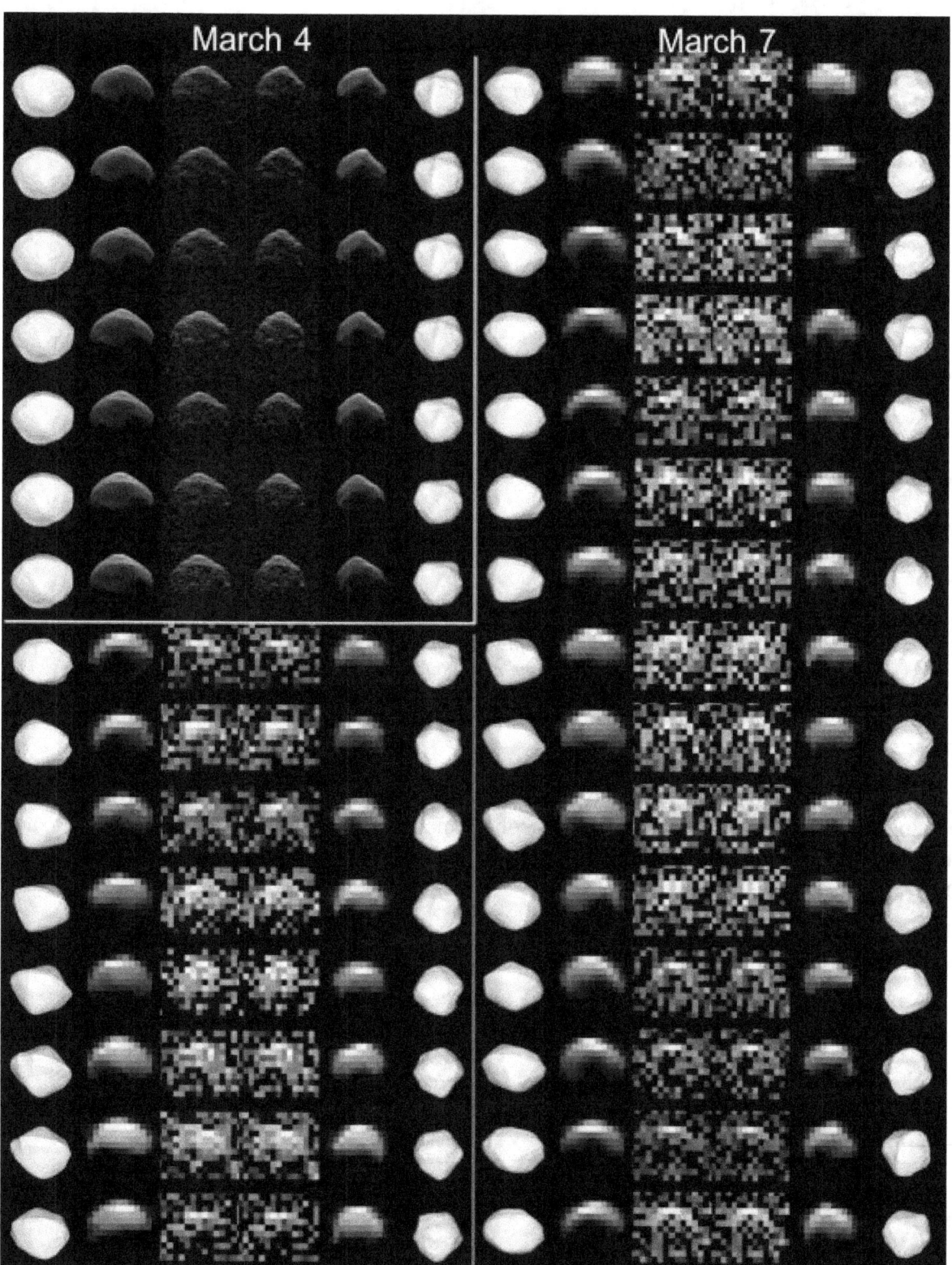

1950 DA image collage, page 4 of 4

Appendix 1.c. 2008 EV5

2008 EV5 Master Log

Time (UTC)	RA,Dec (º)		Dist (AU)	Setup (res) (µs)	# Runs	Subradar Position (long start, end; lat) (º)		
Goldstone								
2008 Dec 16 13:58–14:10	148	–26	0.028	0.125	12	325	341	–28.8
2008 Dec 17 10:00–12:05	147	–23	0.026	0.125	128	200	108	–25.4
2008 Dec 19 08:56–12:15	145	–13	0.024	0.125	195	350	30	–16.0
2008 Dec 21 08:16–13:39	142	–2	0.022	0.125	414	1.45 periods		–5.5
2008 Dec 23 10:24–12:00	139	11	0.022	0.125	131	298	155	+7.4
Arecibo								
2008 Dec 23 06:57–08:19	139	10	0.022	0.05	107	268	144	+6.2
2008 Dec 24 06:25–08:00	138	17	0.022	0.05	108	165	17	+12.5
2008 Dec 26 05:55–08:17	134	29	0.022	0.05	119	257	40	+24.5
2008 Dec 27 06:21–07:41	133	34	0.023	0.05	84	60	298	+30.0
Arecibo+VLBA+GBT								
2008 Dec 23 08:24–08:51	139	10	0.022	CW–multistation	131		91	+6.5

Log of the observations of 2008 EV5 used in shape modeling. CW data are uncoded transmissions, resolving the target only in frequency. The ranging and delay-Doppler data have resolution in time delay (µs), while the VLBA+GBT plane-of-sky data have resolution in angular size (mas). Subradar longitude and latitude are given for our final shape model, with longitude measured relative to the model's +x axis. On 2008 Dec 21, the Goldstone track covered all sub-radar longitudes (more than one complete rotation). All transmissions were left-handed circular polarized.

Following is a collage of the delay-Doppler images of 2008 EV5. Each row of the collage contains an observed delay-Doppler image ('obs'), model images derived from the prograde and retrograde shape models ('pro fit' and 'ret fit'), and plane-of-sky views of the models at that time ('pro sky' and 'ret sky').

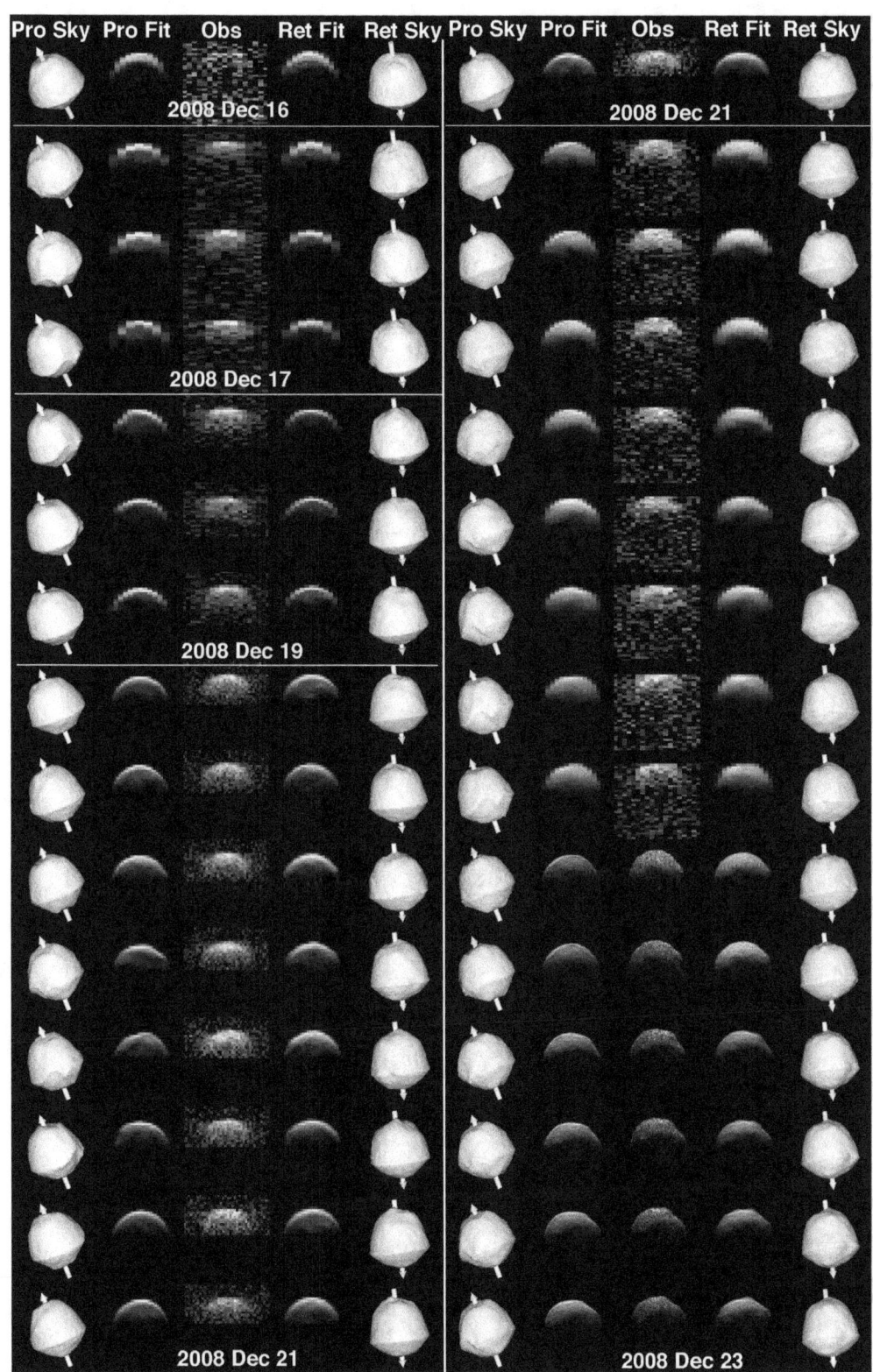

Pro Sky   Pro Fit   Obs   Ret Fit   Ret Sky   Pro Sky   Pro Fit   Obs   Ret Fit   Ret Sky

2008 Dec 16

2008 Dec 21

2008 Dec 17

2008 Dec 19

2008 Dec 21

2008 Dec 23

2008 EV5 image collage, page 1 of 3

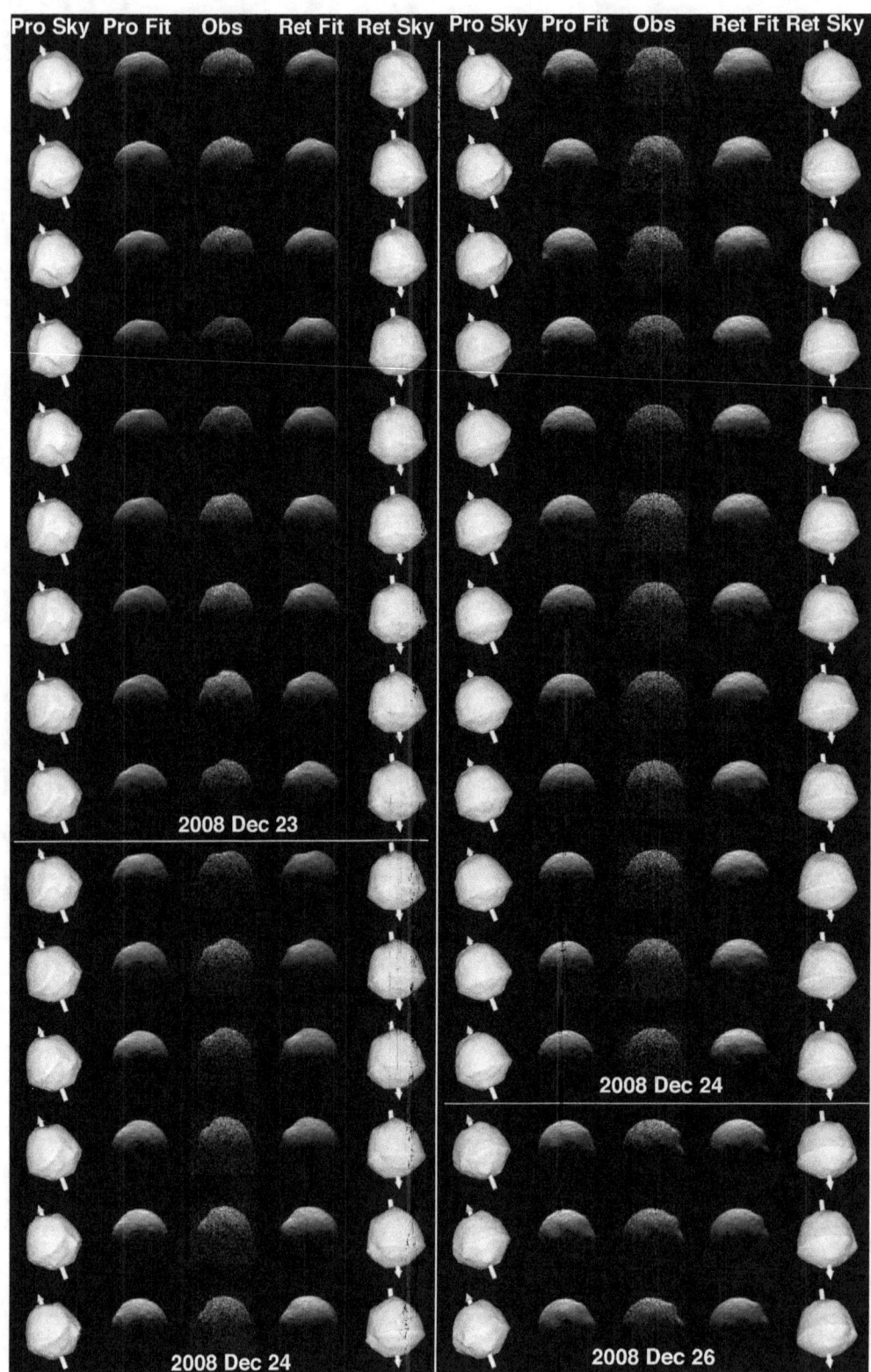

2008 EV5 image collage, 2 of 3

Pro Sky   Pro Fit   Obs   Ret Fit   Ret Sky   Pro Sky   Pro Fit   Obs   Ret Fit   Ret Sky

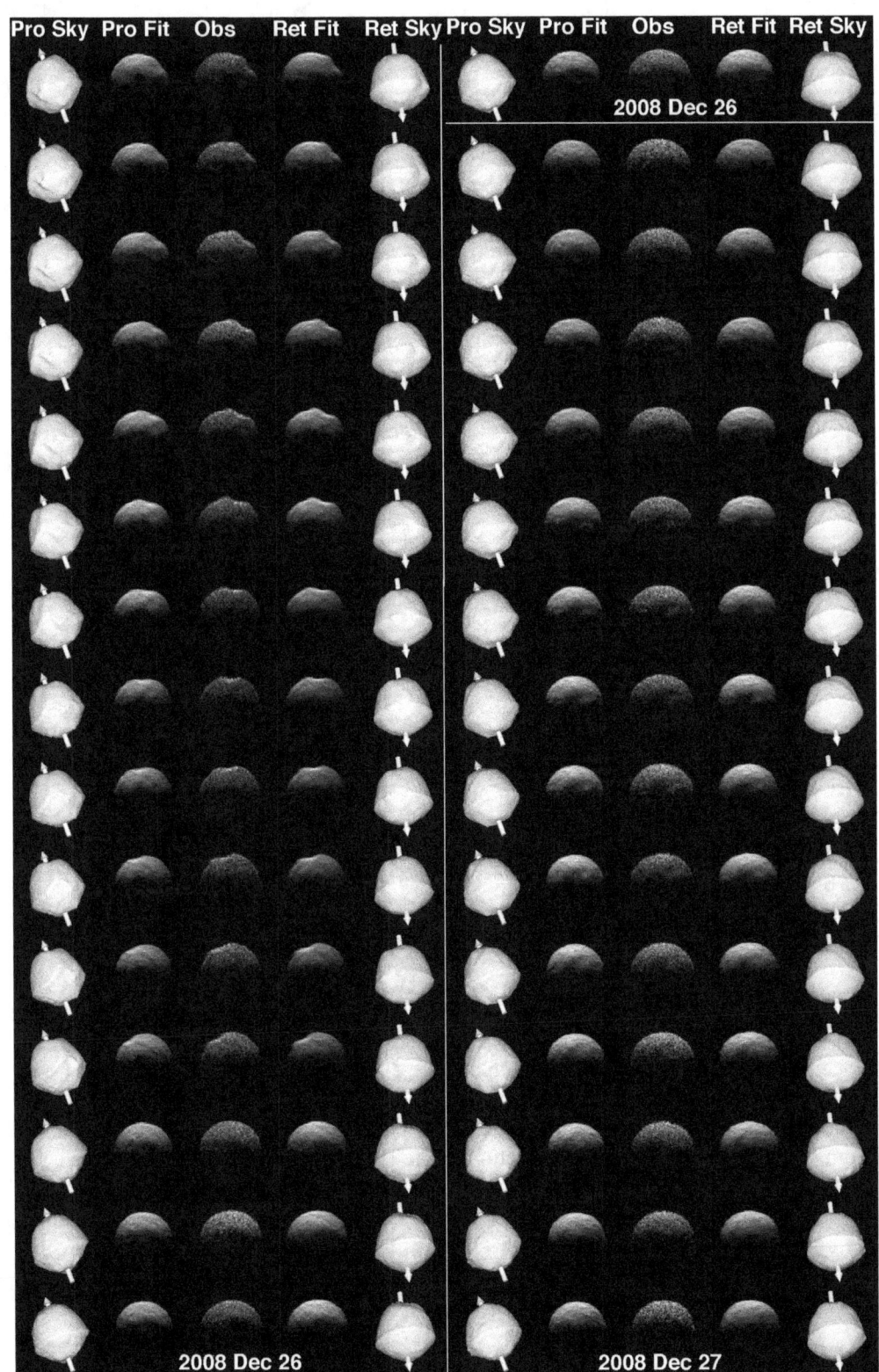

2008 Dec 26

2008 Dec 26

2008 Dec 27

2008 EV5 image collage, page 3 of 3

Appendix 2: Narrow-Band Software Correlator

The following description of my software correlator uses the input, settings, and output of a run on the 2008 EV5 observation on 2008 Dec 23 as examples.

Appendix 2.a: Input Files

The correlator requires the following input files:

Data files from each VLBA station spanning a scan on the target (i.e. each file contains only times when the antenna was pointed at the asteroid). These files must have the format <time> <voltage> <time> <voltage> ..., with time measured in seconds from UT midnight. I obtain these files from the raw VLBA Mark 5 binary data using a portion of the VLBA data-module processing pipeline, copyright 2006 by W. Brisken (available on request). The EV5 data were 2-bit sampled at 8 MHz, and downsampled by a factor of 32 to accelerate the correlation (the filter bandwidth was 62.5 kHz).

```
gb.txt   mk.txt   ov.txt   kp.txt   pt.txt
la.txt   fd.txt   br.txt   nl.txt
```

br.txt
```
        30256.235001     13.343600
        30256.235005     53.374401
        30256.235008     44.030800
        30256.235013     -15.007700
        30256.235016     -48.702599
        30256.235021     -2.335900
        30256.235025     48.702599
        30256.235028     32.351299
        30256.235033     -35.023102
        30256.235037     -53.374401
        ...
```

Ephemeris text files, providing sky position of the target and its Doppler shift and time delay relative to geocenter for each station as a function of time. I obtain these data from the JPL Horizons system. For the near-field corrections to be effective, the overall uncertainties in delay and Doppler should be less than a few microseconds and a hertz, respectively. Delay-Doppler astrometry from single-dish observations is essential. The Doppler and delay files must have the stations in the same order as in the control file.

2008EV5_Dec23.txt   (Time s from UTC midnight. RA and Dec °)

TIME	RA	DEC
30240.00	139.541804	10.139768
30300.00	139.5407231	10.1443279
30360.00	139.5396421	10.1488878

...

2008EV5Doppler.txt   (Relative Doppler shift in Hz)

TIME	GB	MK	OV	
30240	94.52759482	3369.87184	1894.256152	...
30300	81.91835833	3366.337475	1884.335002	...
30360	69.30832796	3362.738011	1874.378921	...

...

2008EV5Delay.txt   (Relative time delay in s)

TIME	GB	MK	OV		
30240	−0.018728076	−0.005717795	−0.015084957	...	
30300	−0.0187303	−0.005802699	−0.015132583	...	
30360	−0.018732206	−0.005887513	−0.015179958	...	...

The calibration file contains complex gain corrections as a function of time for each station. Typically, this file is the output of a more general analysis package such as AIPS, applied to calibrator sources.

gain_estimate.txt   (Complex gains for each station, relative only)

TIME	GB		MK	...	
	REAL	IMAG	REAL	IMAG	...
30240.27758	−2.8558	0.2997	0.7470	18.438	...
30255.27758	−2.8484	0.2990	0.7488	18.481	...
30270.27758	−2.8410	0.2982	0.7506	18.525	...

...

The control text file, 'corr_settings.txt', must be in the same directory as the executable. It includes all of the information necessary for correlation: station positions, data file names, etc. The stations must be ordered here exactly as they are in the ephemeris and gain files (e.g., here Green Bank is given first, followed by Mauna Kea, etc.).

corr_settings.txt
number_of_stations: 9

station_lats:	38.43284	19.80138	37.23165
	31.95631	34.30100	35.77513
	30.63503	48.13123	41.77143
station_lon:	280.1602	204.5445	241.7229
	248.3876	251.8808	253.7544
	256.0552	240.3167	268.4259

station_r:	6365	6359	6365	6363	6364
	6364	6362	6369	6366	

input_data:	../gb.txt	../mk.txt	../ov.txt
	../kp.txt	../pt.txt	../la.txt
	../fd.txt	../br.txt	../nl.txt

```
mjd:                     54823
sample_rate:             250000
carrier_wavelength:      0.126
carrier_offset:          -30000
bandwidth:               5

fft_length:              262144
number_ffts:             100

min_tlag:                -2.5
max_tlag:                2.5
tlag_step:               0.01

doppler_cor:                        2008Dec23Doppler
delay_cor:                          2008Dec23Delay
ephemeris:                          2008EV5_Dec23.txt
ephemeris_lines:                    61
gain_corrections:                   gain_estimate.txt
gain_lines:                         121

vis_output:   20081223_speckletracking.txt
```

Appendix 2.b: Output

The output of the correlator is sent to a file and to the screen. The file is a series of cross-correlations as a function of frequency and time (at the first station in the pair) for each time lag and baseline, covering the user-specified spectral range. Since the stations start recording at slightly different times, each baseline and time lag spans a different interval. The visibilities during a single scan vary due to speckles and noise and gain fluctuations. The screen output is diagnostic information.

```
20081223_speckletracking.txt
Station 0 and 1   Offset: -2.500000 s
30258.761719      545546.871783     15323592.564168 …
30259.287109      20529719.155773 2478982.288627    …
…
Station 0 and 1   Offset: -2.490000 s
30258.771484      341948.795793     16496584.954346 …
…


SCREEN OUTPUT:
>./correlate
     Number of Stations: 9
     Station Latitudes (deg):           38.432838   …
     Station Longitudes (deg):          280.160187 …
```

```
Station Radii (km):      6365 …
Data files:      ../ev5_s01gb.txt …
Modified Julian Day of Observation: 54823
Sampling Rate (Hz): 250000.000000
Wavelength (m): 0.126000
Frequency Offset (Hz): -30000
Final Integration Bandwidth (Hz): 150.000000
FFT Length: 131072
Number of Transforms: 150
Minimum Time Lag (s): -2.500000
Maximum Time Lag (s): 2.500000
Time Lag Step (s): 0.010000
Doppler Ephemeris File: 2008Dec23Doppler
Delay Ephemeris File: 2008Dec23Delay
Sky Position Ephemeris File: 2008EV5_Dec23.txt
Ephemeris Files have 61 lines.
Gain Corrections File: gain_estimate.txt
Gain File has 121 lines.
Visibility Output File: 20081223_speckletracking.txt
Starting Correlation.
Correlating files: ../ev5_s01gb.txt and
../ev5_s01gb.txt, with -2.500000 s offset
Offset: -2.500000  Spectrum:  495901184 …
Offset: -2.400000 …

…
```

All further processing, computing the time lag with maximum cross-correlation for each baseline, is handled outside the code.

Appendix 2.c: Source Code

The executable was compiled from the following source code. The code was developed using gcc 4.0.1 for Mac OS X and has not been thoroughly tested on other systems.

```
/*program to read and reduce VLBA data from text data files*/
/*formatted for gcc compilation on Mac OS 10.X*/
/*Michael Busch*/
/*California Institute of Technology*/
/*busch@caltech.edu*/

#include <stdio.h>
#include <stdlib.h>
#include <math.h>

/*function prototypes*/
  /*utility function - degrees to radians*/
  float d2r(float d) {return d*1.745329252e-2;}
```

```c
/*functions to compute sub lat and sub lon - dec and ra */
/* in rad.  Accurate only to millirad level*/
float sub_lat(double JD, double dec, double ra);
float sub_lon(double JD, double ra);

/*function to compute uv plane positions*/
int uv_compute(float lat_s, float lon_s, int nstat,
               float lat[nstat], float lon[nstat],
               float radius[nstat], float u[nstat][nstat],
               float v[nstat][nstat], float lambda);

/*function to compute FFTs of data*/
int fft(float x[], float y[], float u[], float v[], int length,
        int direction);

/*function to compute corrections for a time array*/
int interpcorrections(char infilename[], float times[],
                      float corrections[], int nlines,
                      int nstations, int npoints, int station);

/*function to apply a constant delay and Doppler offset —
   samplerate in Hz, Doppler Hz, delay s - and compute
   correlation*/
int shift_corr(char infile1[], char infile2[], int stat1,
               int stat2, int nstat, float lat[nstat],
               float lon[nstat], float radius[nstat],
               float u[nstat][nstat], float v[nstat][nstat],
               char dopfile[], char delfile[], char ephfile[],
               char gainfile[], float cor[2], int n,
               int blocks, int mjd, int nlines, int gainlines,
               float samplerate, double offset, float band,
               float lambda, float corba[], char vis_output[],
               double offtime);

/*function to read and interpolate gain corrections*/
int interp_gain_corrections(char infilename[], float times,
                            float real_corr[],
                            float imag_corr[], int nlines,
                            int nstations, int station);

/*function to read and interpolate ephemeris*/
int interp_ephem(char infilename[], float time,
                 float position[2], int nlines);

/*main function*/
int main()
{
  int stat, i, j, k;
  char text_field[40] = {0};

  /*Opening the correlator settings file*/
  FILE *corr_settings;
       corr_settings = fopen("corr_settings.txt","r");
```

```c
/*read in the number of stations*/
int nstat;  fscanf(corr_settings,"%s %d",text_field, &nstat);
            printf("Number of Stations: %d\n",nstat);

/*read in the station positions*/
float station_lats[nstat];
      fscanf(corr_settings,"%s",text_field);
      printf("Station Latitudes (deg):\t");
      for(i = 0; i < nstat; i++)
        {fscanf(corr_settings,"%ff",&station_lats[i]);
          printf("%f\t",station_lats[i]); }
      printf("\n");
float station_longs[nstat];
      fscanf(corr_settings,"%s",text_field);
      printf("Station Longitudes (deg):\t");
      for(i = 0; i < nstat; i++)
        {fscanf(corr_settings,"%ff",&station_longs[i]);
          printf("%f\t",station_longs[i]); }
      printf("\n");
float station_radii[nstat];
      fscanf(corr_settings,"%s",text_field);
      printf("Station Radii (km):\t");
      for(i = 0; i < nstat; i++)
        {fscanf(corr_settings,"%ff",&station_radii[i]);
          printf("%f\t",station_radii[i]); }
      printf("\n");

/*read in the data file names*/
char names[nstat][40]; fscanf(corr_settings,"%s",text_field);
      printf("Data files: \t");
      for(i = 0; i < nstat; i++)
      {
        fscanf(corr_settings,"%s",text_field);
        for(j = 0; j < 40; j++)
          {names[i][j] = text_field[j];}
        printf("%s\t",names[i]);
      }
      printf("\n");
char n1[40], n2[40];

/*read in data properties: date of observation, sampling
  frequency, carrier wavelength and offset from the predicted
  echo frequency*/
int mjd; fscanf(corr_settings,"%s",text_field);
        fscanf(corr_settings,"%df",&mjd);
        printf("Modified Julian Day of Observation:
                %d\n",mjd);
float sample_rate; fscanf(corr_settings, "%s", text_field);
                fscanf(corr_settings,"%ff",&sample_rate);
                printf("Sampling Rate (Hz): %f\n",
                        sample_rate);
float lambda; fscanf(corr_settings, "%s", text_field);
            fscanf(corr_settings,"%ff",&lambda);
            printf("Wavelength (m): %f\n", lambda);
```

```c
double offset;  fscanf(corr_settings, "%s", text_field);
                fscanf(corr_settings,"%lf",&offset);
                printf("Frequency Offset (Hz): %g\n", offset);

/*read in settings for correlation: final bandwidth for whole-
  object visibilities, FFT length and number of transforms*/
float bandwidth;  fscanf(corr_settings, "%s", text_field);
                  fscanf(corr_settings,"%ff",&bandwidth);
                  printf("Final Integration Bandwidth (Hz): 
                          %f\n",bandwidth);
int n; fscanf(corr_settings, "%s", text_field);
       fscanf(corr_settings,"%df",&n);
       printf("FFT Length: %d\n",n);
int blocks; fscanf(corr_settings, "%s", text_field);
            fscanf(corr_settings,"%df",&blocks);
            printf("Number of Transforms: %d\n");

float min_tlag; fscanf(corr_settings, "%s", text_field);
                fscanf(corr_settings,"%ff",&min_tlag);
                printf("Minimum Time Lag (s): 
                        %f\n",min_tlag);
float max_tlag; fscanf(corr_settings, "%s", text_field);
                fscanf(corr_settings,"%ff",&max_tlag);
                printf("Maximum Time Lag (s): 
                        %f\n",max_tlag);
float step_tlag; fscanf(corr_settings, "%s", text_field);
                 fscanf(corr_settings,"%ff",&step_tlag);
                 printf("Time Lag Step (s): %f\n",step_tlag);

/*read in ephemeris and gain file names and number of lines*/
char dopfile[40]; fscanf(corr_settings, "%s", text_field);
                  fscanf(corr_settings,"%s",dopfile);
                  printf("Doppler Ephemeris File: 
                          %s\n",dopfile);
char delfile[40]; fscanf(corr_settings, "%s", text_field);
                  fscanf(corr_settings,"%s",delfile);
                  printf("Delay Ephemeris File: %s\n",delfile);
char ephfile[40]; fscanf(corr_settings, "%s", text_field);
                  fscanf(corr_settings,"%s",ephfile);
                  printf("Sky Position Ephemeris File: %s\n",
                          ephfile);
int ephemeris_lines; fscanf(corr_settings, "%s", text_field);

fscanf(corr_settings,"%df",&ephemeris_lines);
    printf("Ephemeris Files have %d lines.\n",ephemeris_lines);
    char gainfile[40]; fscanf(corr_settings, "%s", text_field);
                       fscanf(corr_settings,"%s",gainfile);
                       printf("Gain Corrections File: %s\n",
                               gainfile);
int gainlines; fscanf(corr_settings, "%s", text_field);
               fscanf(corr_settings,"%df",&gainlines);
               printf("Gain File has %d lines.\n",gainlines);

/*read in the output file names*/
```

```c
char vis_output[40]; fscanf(corr_settings, "%s", text_field);
                     fscanf(corr_settings,"%s",vis_output);
                     printf("Visibility Output File:
                               %s\n",vis_output);

/*other necessary variables*/
float re[nstat][nstat], im[nstat][nstat];
float u[nstat][nstat], v[nstat][nstat];
float gains[nstat];
float cor[2], corba[2], background[nstat][nstat];

double offtime;

/*start the computation*/
printf("Starting Correlation.\n");

/*compute correlation between pairs of antennas*/
for(i = 0; i < n; i++)
{
  for(j = 0; j < n; j++)
  {
    if(i <= j )
    {
      for(offtime = min_tlag;
          offtime <= max_tlag;
          offtime += step_tlag)
      {
       printf("Correlating files: %s and %s, with %f s
               offset\n", names[i], names[j],offtime);

       for(k = 0; k < 40; k++)
         {n1[k] = names[i][k]; n2[k] = names[j][k];}

       stat = shift_corr(n1,n2, i, j, nstat, station_lats,
                          station_longs, station_radii, u, v,
                          dopfile, delfile, ephfile, gainfile,
                          cor, n, blocks, mjd, ephemeris_lines,
                          gainlines, sample_rate, offset,
                          bandwidth, lambda, corba, vis_output,
                          offtime);

       re[i][j] = cor[0]; re[j][i] = cor[0];
       im[i][j] = cor[1]; im[j][i] = -1*cor[1];
       u[j][i] = -1*u[i][j]; v[j][i] = -1*v[i][j];

       background[i][j] = sqrt(corba[0]*corba[0] +
                               corba[1]*corba[1]);
       background[j][i] = background[i][j];
      }
    }
  }
}

/*gain computation*/
```

```c
  printf("\nIntegrated Gains, based on autocorrelations:\n");
  for(i = 0; i < nstat; i++)
  {
    gains[i] = sqrt(re[i][i] - background[i][i]);
    printf("Gain %d: %f\n",i,gains[i]);
  }

  /*normalize*/
  printf("\nNormalized visibilities, integrated over time and
          frequency:\n");
  for(i = 0;  i < nstat; i++)
  {
    for(j = 0; j < nstat; j++)
    {
      re[i][j] = re[i][j]/gains[i]/gains[j];
      im[i][j] = im[i][j]/gains[i]/gains[j];
      printf("Station: %d\tStation: %d\tU: %f\tV: %f\t
             Vr: %f\tVi: %f\n",i,j,u[i][j],v[i][j],
             re[i][j],im[i][j]);
    }
  }

  return 0;
}

/*fft function - x and y are input, u and v transform*/ /*modified from
Numerical Recipes*/
/*packing for inverse transform doesn't work yet*/
int fft(float x[], float y[], float u[], float v[], int length,
        int direction)
{
  float * data = (float*) calloc (2*length+1,sizeof(float));
  int p = 0;

  /*populating array with reals and imaginaries - indexing
    starting with 1*/
  for(p = 0; p < length; p++)
  {
    data[2*p+1] = x[p];
    data[2*p+2] = y[p];
  }

  /*forward transform.  -1 for inverse*/
  int nn = length, isign = direction;

  int n, mmax, m, j, istep, i;
  double wtemp, wr, wpr, wpi, wi, theta; /*double precision*/
  float tempr, tempi, hold;

  /*for inverse transform, pack data for manipulation*/
  if(isign == -1)
  {
    /*data now contains the transform.  Unpacking to u and v*/
    for(p = 0; p < length/2; p++)
```

```
    {
      data[length+2*p+1] = x[p];
      data[length+2*p+2] = v[p];
    }

    for(p = length/2; p < length; p++)
    {
      data[2*p+1 - length] = x[p];
      data[2*p+2 - length] = y[p];
    }

}

/*bit reversal*/
n = nn << 1;
j = 1;
for(i=1; i < n; i += 2)
{
  if(j > i)
  {
      hold = data[j];
      data[j] = data[i]; data[i] = hold;

      hold = data[j+1];
      data[j+1] = data[i+1]; data[i+1] = hold;
  }

  m = n >> 1;
  while(m >= 2 && j > m)
  {
    j -= m;
    m >>= 1;
  }

  j += m;
}

/*Danielson-Lanczos*/
mmax = 2;
while(n > mmax)
{
  istep = 2*mmax;
  theta = 6.28318530717959/(isign*mmax);
  wtemp = sin(0.5*theta);
  wpr = -2.0*wtemp*wtemp;
  wpi = sin(theta);
  wr = 1.0;
  wi = 0.0;

  for(m = 1; m < mmax; m += 2)/*nested inner lopps*/
  {
    for(i = m; i <= n; i += istep)
    {
      j = i + mmax;
```

```
            tempr = wr*data[j] - wi*data[j+1];
            tempi = wr*data[j+1] + wi*data[j];
            data[j] = data[i]-tempr;
            data[j+1] = data[i+1]-tempi;
            data[i] += tempr;
            data[i+1] += tempi;
        }

        wr = (wtemp=wr)*wpr - wi*wpi + wr;
        wi = wi*wpr + wtemp*wpi + wi;
    }

    mmax = istep;
}

/*data now contains the transform.  Unpacking to u and v*/
for(p = 0; p < length/2; p++)
{
    u[p] = data[length+2*p+1];
    v[p] = data[length+2*p+2];
}

for(p = length/2; p < length; p++)
{
    u[p] = data[2*p+1 - length];
    v[p] = data[2*p+2 - length];
}

free(data);

return 0;
}

/*function to apply a constant delay and Doppler offset —
   samplerate in Hz, Doppler Hz, delay s - and compute the
   correlation*/
int shift_corr(char infile1[], char infile2[], int stat1,
                int stat2, int nstat, float lat[nstat],
                float lon[nstat], float radius[nstat],
                float u[nstat][nstat], float v[nstat][nstat],
                char dopfile[], char delfile[], char ephfile[],
                char gainfile[], float cor[2], int n, int blocks,
                int mjd, int nlines, int gainlines,
                float samplerate, double offset, float band,
                float lambda, float corba[], char vis_output[],
                double offtime)
{
    /*opening files for reading*/
    FILE *ifile1; ifile1 = fopen(infile1,"r");
    FILE *ifile2; ifile2 = fopen(infile2,"r");

    float * times1 = (float*) calloc (n,sizeof(float));
    float * times2 = (float*) calloc (n,sizeof(float));
    float * delaycor1 = (float*) calloc (n,sizeof(float));
```

```c
float * delaycor2 = (float*) calloc (n,sizeof(float));
float * dopcor1 = (float*) calloc (n,sizeof(float));
float * dopcor2 = (float*) calloc (n,sizeof(float));
float * in1 = (float*) calloc (n,sizeof(float));
float * in2 =(float*) calloc (n,sizeof(float));
float * real1 = (float*) calloc (n,sizeof(float));
float * imag1 = (float*) calloc (n,sizeof(float));
float * real2 = (float*) calloc (n,sizeof(float));
float * imag2 = (float*) calloc (n,sizeof(float));
float * t1r = (float*) calloc (n,sizeof(float));
float * t1i = (float*) calloc (n,sizeof(float));
float * t2r = (float*) calloc (n,sizeof(float));
float * t2i = (float*) calloc (n,sizeof(float));
float gain1r[1];
float gain1i[1];
float gain2r[1];
float gain2i[1];
float freq, res = (float) samplerate/n;
int i, j, stat, s1, s2;
double time1, time2;
double delay1, delay2;
int diff, numblocks = 1;
float tinit;
float cori, corr, core0, core1;
float corb0, corb1;
float * cori_h = (float*) calloc (n,sizeof(float));
float * corr_h = (float*) calloc (n,sizeof(float));
float lat_s, lon_s;
float u_temp[nstat][nstat], v_temp[nstat][nstat];
float position[2];
double i_float;

FILE *visibilityfile = fopen(vis_output,"a");
fprintf(visibilityfile,"Offset: %f\n",offtime);

offset = 6.2831853071796*offset/samplerate;

fscanf(ifile1,"%lf %ff",&time1,&in1[0]);
fscanf(ifile2,"%lf %ff",&time2,&in2[0]);

/*synching times*/
if(time1 <= time2) { tinit = time1;} else {tinit = time2;}

i_float = time1;
  while( i_float < tinit + offtime)
    {fscanf(ifile1, "%lf %ff", &time1, &in1[0]);
     i_float = time1;}
i_float = time2;
  while( i_float < tinit)
    {fscanf(ifile2, "%lf %ff", &time2, &in1[0]);
     i_float = time2;}

cor[0] = 0; cor[1] = 0;
for(i = 0; i < n; i++) {cori_h[i] = 0; corr_h[i] = 0;}
```

```c
corba[0] = 0; corba[1] = 0;

for(i = 0; i < blocks; i++)
{
  for(j = 0; j < n; j++)
  {
    fscanf(ifile1,"%f %ff",&times1[j],&in1[j]);
    fscanf(ifile2,"%f %ff",&times2[j],&in2[j]);
  }

  /*reading and computing corrections*/
  stat = interpcorrections(dopfile,times1, dopcor1, nlines,
                           nstat, n, stat1);
  stat = interpcorrections(dopfile,times2, dopcor2, nlines,
                           nstat, n, stat2);
  stat = interpcorrections(delfile,times1, delaycor1, nlines,
                           nstat, n, stat1);
  stat = interpcorrections(delfile,times2, delaycor2, nlines,
                           nstat, n, stat2);

  /*initializing correlation*/
  corr = 0; cori = 0; core0 = 0; core1 = 0;

  numblocks++;

  /*taking out carrier offset and Doppler shift*/
  for(j = 0; j < n; j++)
  {
    real1[j] = in1[j]*cos((dopcor1[j]*-6.2831853071796/
                          samplerate + offset)*j);
    imag1[j] = in1[j]*sin((dopcor1[j]*-6.2831853071796/
                          samplerate + offset)*j);
    real2[j] = in2[j]*cos((dopcor2[j]*-6.2831853071796/
                          samplerate + offset)*j);
    imag2[j] = in2[j]*sin((dopcor2[j]*-6.2831853071796/
                          samplerate + offset)*j);
  }

  /*compute fft*/
  stat =  fft(real1, imag1, t1r, t1i, n, 1);
  stat = fft(real2, imag2, t2r, t2i, n, 1);

  /*swap back*/
  for(j = 0; j < n; j++)
  {
    real1[j] = t1r[j]; imag1[j] = t1i[j];
    real2[j] = t2r[j]; imag2[j] = t2i[j];
  }

  delay1 = 0; delay2 = 0;

  /*reading and computing gain corrections*/
  stat = interp_gain_corrections(gainfile, times1[n/2],
                                 gain1r, gain1i, gainlines,
```

```
                              nstat, stat1);

stat = interp_gain_corrections(gainfile, times2[n/2],
                               gain2r, gain2i, gainlines,
                               nstat, stat2);

/*delay corrections by phase rotation and gain corrections by
  multiplication*/
for(j = 0; j < n; j++)
{
   delay1 += -6.2831853071796*delaycor1[j]/samplerate;
   t1r[j] = real1[j]*cos(delay1) - imag1[j]*sin(delay1);
   t1i[j] = imag1[j]*cos(delay1) + real1[j]*sin(delay1);

   delay2 += -6.2831853071796*delaycor2[j]/samplerate;
   t2r[j] = real2[j]*cos(delay2) - imag2[j]*sin(delay2);
   t2i[j] = imag2[j]*cos(delay2) + real2[j]*sin(delay2);

   real1[j] = (t1r[j]*gain1r[0] + t1i[j]*gain1i[0])/
              (gain1r[0]*gain1r[0] + gain1i[0]*gain1i[0]);
   imag1[j] = (t1i[j]*gain1r[0] - t1r[j]*gain1i[0])/
              (gain1r[0]*gain1r[0] + gain1i[0]*gain1i[0]);

   real2[j] = (t2r[j]*gain2r[0] + t2i[j]*gain2i[0])/
              (gain2r[0]*gain2r[0] + gain2i[0]*gain2i[0]);
   imag2[j] = (t2i[j]*gain2r[0] - t2r[j]*gain2i[0])/
              (gain2r[0]*gain2r[0] + gain2i[0]*gain2i[0]);
}

for(j = 0; j < n; j++) {corr_h[j] += real1[j]*real2[j] +
                                     imag1[j]*imag2[j];
                        cori_h[j] += imag1[j]*real2[j] -
                                     real1[j]*imag2[j];}

corr = 0; cori = 0;

for(j = (n - ((int) (band/res)))/2;
    j <= (n + ((int) (band/res)))/2;
    j++)
{
   corr += real1[j]*real2[j] + imag1[j]*imag2[j];
   cori += imag1[j]*real2[j] - real1[j]*imag2[j];
}

cor[0] += corr; cor[1] += cori;

corb0 = 0; corb1 = 0;

corba[0] += corb0; corba[1] += corb1;

core0 = 0; core1 = 0;

/*computing _approximate_ sub lat and lon*/
/*2008 EV5*/
```

```c
        stat = interp_ephem(ephfile, times1[n/2], position, nlines);

        lat_s = sub_lat(2400000.5 + mjd + times1[n/2]*1.1574074e-5,
                        d2r(position[1]), d2r(position[0]));
        lon_s = sub_lon(2400000.5 + mjd + times1[n/2]*1.1574074e-5,
                        d2r(position[0]));

        /*computing approximate uv plane positions*/
        stat = uv_compute(lat_s, lon_s, nstat,  lat, lon, radius,
                        u_temp, v_temp, lambda);
        /*saving UV position at mid-epoch*/
        if(i == ((int) blocks*0.5) )
            { u[stat1][stat2] = u_temp[stat1][stat2];
              v[stat1][stat2] = v_temp[stat1][stat2];}

        /*output amplitude at each frequency at each block*/
        fprintf(visibilityfile,"%f\t",times1[n/2]);

        for(j = n/2 - 20; j <= n/2 + 20; j++)
        {
            core0 = real1[j]*real2[j] + imag1[j]*imag2[j];
            core1 = -1*real1[j]*imag2[j] + real2[j]*imag1[j];
            fprintf(visibilityfile,"%f\t",
                    sqrt(core0*core0+core1*core1));
        }

        fprintf(visibilityfile,"\n");
    }

    /*Output visibilities integrated in time across bandpass*/
        printf("Offset: %f\tSpectrum:\t",offtime);

        for(j = n/2 - 20; j < n/2 + 20; j++)
        {
            core0 = sqrt(corr_h[j]*corr_h[j]+cori_h[j]*cori_h[j]);
            printf("%f\t",core0);
        }

        printf("\n");

    /*closing files and freeing up memory*/
    fclose(ifile1); fclose(ifile2); fclose(visibilityfile);
    free(in1); free(in2); free(t1r); free(real1); free(imag1);
    free(real2); free(imag2); free(t1i); free(t2r); free(t2i);
    free(corr_h); free(cori_h);

    return 0;
}

/*function to compute the sub_lat - dec and ra in rad*/
float sub_lat(double JD, double dec, double ra)
{
    float lat;
```

```
                 /*second term is precession offset to first order*/
                 lat = dec + 9.696e-5*(JD - 2451545.5)/365.2422*cos(ra);
                 return lat;
}

/*function to compute the sub_lon - dec and ra in rad*/
float sub_lon(double JD, double ra)
{
    float lon;
    double d, gmst, f;

    d = JD - 2451545.0;

    ;/*computation of gmst, errors 0.1 s per century*/
    gmst = 18.697374558 + 24.06570982441908*d;

    gmst = (gmst/24); gmst = gmst - floor(gmst);
    gmst = gmst*24;/*taking out phase*/

    lon = ra - gmst*0.261799;/*converting from hours to radians*/

    if(lon < 0) {lon = lon + 6.2831853072;}

    return lon;
}

/*function to compute uv plane positions.*/
/*all angles must be given in radians, r in km from geocenter,*/
/* and lambda in m*/
int uv_compute(float lat_s, float lon_s, int nstat,
               float lat[nstat], float lon[nstat],
               float radius[nstat], float u[nstat][nstat],
               float v[nstat][nstat], float lambda)
{
    float x[9], y[9];
    float s1, s2, s3;
    int i, j;
    float lat_temp, lon_temp;

    /*compute station position vector and project to normal plane*/
    for(i = 0; i < nstat; i++)
    {
        lat_temp = d2r(lat[i]); lon_temp = d2r(lon[i]);

        s1 = radius[i]*cos(lon_temp)*cos(lat_temp);
        s2 = radius[i]*sin(lon_temp)*cos(lat_temp);
        s3 = radius[i]*sin(lat_temp);

        x[i] = -1*sin(lon_s)*s1 + cos(lon_s)*s2;
        y[i] = -1*cos(lon_s)*sin(lat_s)*s1
               - sin(lon_s)*sin(lat_s)*s2 + cos(lat_s)*s3;
    }

    /*compute uv plane positions*/
```

```
   for(i = 0; i < nstat; i++)
   {
     for(j = 0; j < nstat; j++)
     {
        if(i == j) {u[i][j] = 0; v[i][j] = 0;}
        else
        {
             u[i][j] = (x[i] - x[j])*1000/lambda;
             v[i][j] = (y[i] - y[j])*1000/lambda;
        }
     }
   }

   return 0;
}

/*function to compute corrections for a time array*/
int interpcorrections(char infilename[], float times[],
                      float corrections[], int nlines,
                      int nstations, int npoints, int station)
{
  FILE *infile;
  char names[40];
  float intimes[nlines];
  float incorrections[nstations][nlines];
  int i, j;
  double frac;

  infile = fopen(infilename,"r");

  fscanf(infile,"%sf",&names);

  for(i = 0; i < nstations; i++) {fscanf(infile,"%sf",&names);}

  for(i = 0; i < nlines; i++)
  {
    fscanf(infile,"%ff",&intimes[i]);

    for(j = 0; j < nstations; j++)
    {
      fscanf(infile,"%ff",&incorrections[j][i]);
    }
  }

  fclose(infile);

  if(station >= nstations || station < 0)
    {printf("Bad station number\n");}
  if(times[0] < intimes[0] ||
     times[npoints-1] > intimes[nlines-1])
    {printf("Bad time range.  Input time: %f %f \t
            Min time: %f \t Max time: %f\n",times[0],
            times[npoints-1],intimes[0],intimes[nlines-1]);}
```

```c
    for(i = 0; i < npoints; i++)
    {
      j = 0;

      while(intimes[j] < times[i])
      {
        j++;
      }

      frac = (times[i] - intimes[j-1])/(intimes[j]-intimes[j-1]);

      corrections[i] = incorrections[station][j-1]*(1-frac) +
                       incorrections[station][j]*frac;
    }

  return 0;
}

/*function to compute corrections for a time array*/
int interp_gain_corrections(char infilename[], float times,
                            float real_corr[], float imag_corr[],
                            int nlines, int nstations,
                            int station)
{
  FILE *infile;
  char names[40];
  float intimes[nlines];
  float real_incorrections[nstations][nlines];
  float imag_incorrections[nstations][nlines];
  int i, j;
  double frac;
  float real_bp[nstations];
  float imag_bp[nstations];
  float temp_r, temp_i;

  /*reading in bandpass corrections table*/
  for(i = 0; i < nstations; i++)
  {
    real_bp[i] = 1;
    imag_bp[i] = 0;
  }

  /*reading in CL corrections table*/
  infile = fopen(infilename,"r");

  fscanf(infile,"%sf",&names);

  /*reading in header lines*/
  for(i = 0; i < 3*nstations; i++) {fscanf(infile,"%sf",&names);}

  for(i = 0; i < nlines; i++)
  {
    fscanf(infile,"%ff",&intimes[i]);
```

```c
  for(j = 0; j < nstations; j++)
  {
    /*reading in gains*/
    fscanf(infile,"%ff",&real_incorrections[j][i]);
    fscanf(infile,"%ff",&imag_incorrections[j][i]);
  }
}

fclose(infile);

if(station >= nstations || station < 0)
  {printf("Bad station number\n");}
if(times < intimes[0] || times > intimes[nlines-1])
  {printf("Bad time range: Input time: %f \t Min time: %f \t
          Max time: %f\n",times,intimes[0],
          intimes[nlines-1]);}

  j = 0;

  while(intimes[j] < times)
  {
    j++;
  }

  frac = (times - intimes[j-1])/(intimes[j]-intimes[j-1]);

  real_corr[0] = real_incorrections[station][j-1]*(1-frac)
               + real_incorrections[station][j]*frac;
  imag_corr[0] = imag_incorrections[station][j-1]*(1-frac)
               + imag_incorrections[station][j]*frac;

  temp_r = real_corr[0]*real_bp[station]
         - imag_corr[0]*imag_bp[station];
  temp_i = imag_corr[0]*real_bp[station]
         + real_corr[0]*imag_bp[station];

  real_corr[0] = temp_r; imag_corr[0] = temp_i;

  return 0;
}

/*function to interpolate ephemeris information*/
int interp_ephem(char infilename[], float time,
                 float position[2], int nlines)
{
  FILE *infile;
  char names[30];
  float intimes[nlines];
  float incorrections[nlines][2];
  int i, j;
  double frac;

  infile = fopen(infilename,"r");
```

```c
fscanf(infile,"%sf",&names); fscanf(infile,"%sf",&names);
fscanf(infile,"%sf",&names);

for(i = 0; i < nlines; i++)
{
  /*reading in times and ephemeris positions*/
  fscanf(infile,"%ff",&intimes[i]);
  fscanf(infile,"%ff",&incorrections[i][0]);
  fscanf(infile,"%ff",&incorrections[i][1]);
}

fclose(infile);

if(time < intimes[0] || time > intimes[nlines-1])
  {printf("Bad time range\n");}

j = 0;

while(intimes[j] < time)
{
  j++;
}

frac = (time - intimes[j-1])/(intimes[j]-intimes[j-1]);

position[0] = incorrections[j-1][0]*(1-frac)
             + incorrections[j][0]*frac;
position[1] = incorrections[j-1][1]*(1-frac)
             + incorrections[j][1]*frac;

return 0;
}
```